THE
SCIENCE
OF
BASEBALL

THE MATH, TECHNOLOGY, AND DATA BEHIND THE GREAT AMERICAN PASTIME

WILL CARROLL
FOREWORD BY PETER GAMMONS

Skyhorse Publishing

Skyhorse Publishing books may be purchased in bulk at special discounts
for sales promotion, corporate gifts, fund-raising, or educational purposes.
Special editions can also be created to specifications. For details, contact the
Special Sales Department, Skyhorse Publishing, 307 West 36th Street,
11th Floor, New York, NY 10018 or info@skyhorsepublishing.com.

Skyhorse® and Skyhorse Publishing® are registered trademarks of
Skyhorse Publishing, Inc.®, a Delaware corporation.

Visit our website at www.skyhorsepublishing.com.

10 9 8 7 6 5 4 3 2 1

Library of Congress Cataloging-in-Publication Data is available on file.

Cover design by David Ter-Avanesyan
Cover images: baseball background by Getty Images;
math overlay by Shutterstock

ISBN: 978-1-5107-6897-0
Ebook ISBN: 978-1-5107-6898-7

Printed in the United States of America

CONTENTS

CONTENTS

FOREWORD BY PETER GAMMONS

I don't remember when I became aware of Will Carroll. He said I mentioned him on the Tony Kornheiser Show on ESPN Radio around 2002, which was before he began writing for Baseball Prospectus. Anyone who has ever covered sports understands that injuries and all health issues have been a major part of the individual and team performances that the media and fans care about, and when we hear casual mention of oblique pulls, ulnar neuritis, or slap tears, we liberal arts majors have no idea of the long-term ramifications of those injuries, or, as is often the case, what part of the anatomy is impacted.

So, whenever I first read "Under The Knife," Carroll's column about baseball injuries, I always seemed to learn something from it.

We are all prisoners of the mysteries, as teams from Major League Baseball to the National Football League to Southeastern Conference Football refuse to discuss or even acknowledge injuries. Some of it is that most managers and coaches believe there is an advantage in opposing teams not knowing the extent of individual injuries. There certainly was a strategy and element of surprise in the first game of the World Series when Kirk Gibson hobbled out of the Dodger dugout to bat against Dennis Eckersley, and thirty-three years later up the California Coast the Dodgers had given mixed signals about whether or not Max Scherzer could come out of the bullpen to try to finish off the deciding game of the National League Division Series. The bullpen gate opened, and Scherzer came out as if he were auditioning for the Running of the Bulls and finished the bottom of the 9th inning.

We see players leave games, and teams often issue updates referring to "upper body" or "lower body" injuries, with general timeframes for whatever the injury must entail. Then when my eyes and mind were opened to Carroll, two things immediately came to mind: Will understands medical science, orthopedics, osteoporosis, and how injuries apply to what a Ted Williams or Joey Votto understands is the supply line connection from the foot though the hips and core to the fingertips in hitting a baseball, or everything involved in the kinetic chain in throwing a baseball.

In other words, he is a man who has studied, listened, and learned a complex field. But equally as important, Carroll is a journalist who is not only an excellent reporter, but for the two decades he has been my guide, he has built a combination of respect for his work and trust from the doctors, trainers, and rehabilitation specialists and doctors to avoid general speculation and know what the athlete is doing to address the injury. For instance, in the 2021 NLDS between the Giants and Dodgers, there was speculation that injured Giants first baseman Brandon Belt might make a dramatic Kirk Gibson appearance at some critical point in the series.

Carroll wrote Belt "continues to do some baseball activities, like fielding grounders in front of everybody…that's not the issue for Belt. Holding a bat, now that's an issue, and we still haven't seen him do it." With the win-or-go-home game down to its final out and Scherzer due to pitch to Wilmer Flores—who had never had a hit against Scherzer—we knew from Carroll that there was no Kirk Gibson alternative, or drama.

When a Noah Syndegaard or Carlos Carrasco is rehabbing, Carroll has the contacts to present the stage and the predictable timetable for their returns. During the 2021 season, when the rehabs of Chris Sale, Jacob deGrom, and Yu Darvish were significant factors, not only was Carroll's information detailed and accurate, but for the rest of us still wondering the whys and whynots, he translated.

Coming from one whose first full season covering Major League Baseball for the *Boston Globe* was 1972, and hence covered the first labor strike four weeks into my first spring training in scenic

Winter Haven, Florida, Carroll's work over two decades has been a resource, if not a kind of sourcebook because of his scholarship, his contacts, and the trust he has earned in getting information. *The Science of Baseball* involves science that fascinates us all—why curveballs break, why grips and fingers determine fastball movement. About fifteen years ago Manny Ramirez was describing why he felt out of sync when he swung and was in a slump, and when I asked, "Are you trying to refigure the connection of your left foot to the fingertips on your right hand?" he replied, "Exactly," and Carroll can explain that.

We hear about the technology that is so prevalent in baseball, in pitch design and tracking, in the Rapsodo and Edgertronic technology as well as the video used to allow application of the appropriate biomechanics to individual bodies to create the most efficient athletic movement. Carroll allows us to understand the physics, technology, and biomechanics that go into why Max Fried is so successful a pitcher, why Walker Buehler could have Tommy John Surgery and *then* embark on a brilliant career, and how Juan Soto's eyesight and inherent tracking system lets him know what pitch is being thrown as it is released.

There was a time when players and coaches liked to simplify hitting down to "see the ball, hit the ball." But it never was simple. Don Mattingly has always said, "Ted Williams and Barry Bonds saw things no one else could see." That is fascinating. So is watching elite hitters two hours before a game taking batting practice with a small camera or even a phone camera attached to the netting of the home plate cage, and how a meticulous hitting student like J.D. Martinez may feel something in his swing is awry, takes the camera in, and studies the video in the indoor cages.

When I first pulled into Winter Haven, Florida, in March 1972, as someone three years and a month into my tenure at the *Boston Globe* to begin my professional life, it was a glorious day, even if it brought me to the shore of Lake Lulu. The culture, demographics, revenue, media, and venues that once seemed fresh now seem primitive.

A wonderful baseball player and lifer named Jim Fregosi gave this advice to a young scout thirty years after that: "If you don't love the game of baseball and love players, find another profession first thing in the morning." That is the basis of why people like Will Carroll and I go to the park hoping to learn two things we did not know.

But one of the joys of watching baseball evolve is that now we better understand why things happen.

When I was a small child, my parents took me to the movies to see *It Happens Every Spring*, a film about a college professor and a baseball hit through a window into his lab that got soaked in some fictional fluid which made the ball allergic to wood and led the professor to take up pitching, utilizing the chemical to throw balls wood bats repelled and pitching the St. Louis Browns into the World Series. As kids, we fantasized about doing the same thing. Today, of course, this would be a Spider Tack violation complete with fines and suspensions.

It Happens Every Spring was about the fantasy of baseball, produced and filmed in Hollywood. Twenty years later, Ted Williams's *The Science of Hitting* was published, and myth began morphing into science. Years later, Harvey Dorfman's *The Mental Game of Baseball* was published, and the humanity and psychology involved with the game began to be understood to the point today when mental skills coaches are in the sport's grains.

Now we have *The Science of Baseball*, which presents important findings regarding the evolution of what actually makes the game of baseball work, and helps highlight how Will Carroll continues to expand our understanding of and fascination for baseball.

This one's for my dad.
"Wanna have a catch?"
– Ray Kinsella

INTRODUCTION

Baseball is a game of comparisons. Old timers love to tell stories and scouts speak in similars, where this guy reminds them of that guy, where this pitcher could be as good as that pitcher. By writing a book with the title, *The Science of Baseball*, I know I'm tempting the Baseball Gods, shelved somewhere between Ted Williams's *The Science of Hitting* and Robert Adair's *The Physics of Baseball*. I know I'll end up like one of those players drafted ahead of Mike Trout, but a couple of them made the big leagues. You might not remember their names, but those names are written in the indelible ink of the chosen few that have played in the big leagues. That's pretty good and for me, writing this and you taking your valuable time to read it is as close to the bigs as I'll get.

No writer (and every writer) approaches a book project hoping to write a classic, but my hope is that in the modern, changing game of baseball, we can still learn to look at the game differently, smarter, and with an eye to making everything better because we understand it more completely. We have that chance in this era more than any before, simply because our technology and our ability to see and understand what is happening is clearer than any time before. If there's any one thing about the game of baseball that I would change, it's that it doesn't change. Whether it's the hidebound traditionalists, or the committee-strewn road to some sort of false consensus, a race between baseball as a sport and a glacier is even odds.

Look at a pitch in 2022 and you'll get all sorts of information, from velocity to break, from location to type. It will be categorized by three systems at once, all working both in concert and against each other. A pitch from not so long ago, say a fastball from Nolan Ryan in one of his no-hitters, and we have almost none of that information. Velocity was a myth for the greater part of the history of baseball, with radar guns rare at the start of Ryan's career and ubiquitous by the end.

This of course will change, as will all things. Putting something in ink is an act of hubris, knowing that from the moment of print, things will change as surely as the seasons. This will be science and technology as exists now, though I do have a chapter that predicts where this will go in both the short and long term. When I look back at some of my writings from 2004, when I wrote *Saving the Pitcher*, I'm still proud of it, but there's so much that's out of date. There were times in this book where I considered leaving something out, but I believe that a book is a stake in time. This is the best I could do now and while I'm sure people will look back and say "Really?," marking that point will show progress, and that's a positive thing.

My goal is to answer some questions and reconfigure how you think and watch the game of baseball. My hope is that you approach it the same way I did—knowing some about baseball, wishing to know more, and remaining open-minded enough to learn. I'm lucky enough to have this chance, and after twenty-plus years in this business, I've met some great people who were willing to share their time and knowledge with me, and now, with you.

Baseball is often a game of statistics, but also stories. It's a game of science, but also of magic. I've tried to balance it, never losing sight that it's men playing the game, and that there's men and women out there surrounding it, while all the time there are children discovering that same magic that we all carry, that love for baseball, that love for truth, and I hope the two are often one and the same.

This isn't a novel and there's no need for you to start at the beginning and read straight through, though you are certainly welcome to do so. The best stories are the ones you know by heart and the best friends are the ones that know the details between our stories. Here are some of mine for you, new friends and old.

Throughout this book, I will reference things that are simply easier to see than fully describe. Because we don't yet have Harry Potter–style books, I have put up a special page at my site—undertheknife.substack.com—that will show these pictures and videos.

1

THE BALL

We call it baseball, but have you ever really looked at—or inside—a baseball? I think one of those magic moments of childhood is when you plaster a ball around the sandlot enough that the stitches come loose and suddenly miles of yarn come out. Someone starts unraveling it and suddenly, someone grabs the end and runs. He's impossibly far and then there's another bundle, before you finally get to the center, Tootsie Pop-like, and there's just this tiny other ball. Do you unravel it? Does it bounce?

The issue of the ball begins in the definition of the object. The rules that govern baseball are often quite detailed. A field is supposed to point east-northeast, looking from home plate to the pitcher and beyond to second base. The definition and distances of the field are laid out to the inch. For the ball, the namesake of the game, there is a precious little said.

Rule 3.01: "The ball shall be a sphere formed by yarn wound around a small core of cork, rubber or similar material, covered with two strips of white horsehide or cowhide, tightly stitched together. It shall weigh not less than five nor more than 5 1/4 ounces avoirdupois and measure not less than nine nor more than 9 1/4 inches in circumference."

Two sentences are all the ball gets and pretty much everything else is left to chance, as opposed to the architectural drawings they include for the mound or the detailed and extensive instructions for the construction of a bat. (More on that later.) This leaves a lot of room for variation and likely made sense when the rule was first put in place. The balls were at that stage almost handmade, but you would likely be surprised to find out that today, they are still largely handmade and therefore inconsistent. That's a problem since much of the basis of the game is that at the very least, a baseball is a baseball is a baseball. Ever see a pitcher get a new ball from the ump, look at it, and toss it right back? It's because they're not the same at all. Let's look backward, from how the ball shows up on field to how it's made.

Going back to 1876 and A.G. Spalding's selection by the National League to become the standard baseball, the ball itself has been largely handmade. There have been changes to the ball, such as the "pill" at the center going from rubber to cork and the surface of the ball changing from horsehide to cowhide in 1974. The biggest change came in 1920, when the death of Ray Chapman led to more frequent changes of the ball. New balls were more lively, filled with an Australian wool, leading to an increase in home runs.

Another major change came in 1974, when MLB went from its historic Spalding-made ball to one built by Rawlings. There was very little change in the actual construction, and the way the game stayed the same in those initial years suggests that the ball was largely the same. The construction was the same, at least in technique and result.

While Rawlings initially produced the ball in Haiti, production shifted to Costa Rica, where it exists today. Rawlings is co-owned by Major League Baseball and Seidler Equity Partners, the majority owners of the San Diego Padres. (The Seidler brothers are the grandsons of Walter O'Malley, who brought the Dodgers to Los Angeles.) The league and the private equity firm teamed up to buy the manufacturer in 2018, the year after the ball first became noted as a big variable.

This creates a number of issues, but let's look first at how the ball is prepared. A key part of the baseball as it comes into the game is mud. Yes, mud. In a book about science, it's somewhat laughable that something like mud from a specific spot could be a fundamental part of the key piece of equipment in a professional sport, but here we are.

The balls are "rubbed up" by umpires using a substance known as "Lena Blackburne's Rubbing Mud." This comes, to this day, from a secret location somewhere along the Delaware River, on the New Jersey side. It's cleaned and screened, then put in jars. One of the few times this has ever been shown was on the television show "Dirty Jobs."

As an aside, Lena Blackburne is not a woman. Russell Aubrey Blackburne was a pitcher then manager for several MLB teams. He marketed his mud as a side job while he coached on Connie Mack's staff, and it became the MLB standard for both leagues in 1938. Prior to that, everything from tobacco juice to infield dirt was used to rough up the balls, a practice still seen today at some college and high school levels. The name? It was a New England adjustment of the nickname "Leaner," given to him because of his rail-thin physique.

To this day, Blackburne's mud, from the same location, is used to rub the sheen off the ball and make it easier to grip. A four-pound container is available to the public for a hundred bucks.

In Japan's NPB (the Japanese top league), the ball is made by Mizuno, with the sheen already off the ball and a bit sticky right out of the box. For the 2020 Olympics, the company that manufacturs them—SSK, from Sri Lanka—used Lena Blackburne's Mud in their process at the Olympics.

That leads many to think that the issue is not the mud, but the leather that the ball is made of. Rawlings is supplied by Horween Leather, a well-known tanner from Chicago. Horween leather is widely sought in applications like watch bands, shoes, and other fine goods. Actually, Horween has been doing baseballs for longer than Rawlings; they were the leather supplier to Spalding as well, prior to the manufacturer switch. While Horween refused comment, I was told that the only change to their processes in years has been the switch from horsehide to cowhide, which happened in 1974.

Of course, the players complained. Hank Aaron himself said the balls didn't carry, even in batting practice. Dick Allen, the White Sox slugger, was more specific, saying the balls seemed smaller and harder, which is the opposite of what most would expect for a ball that didn't seem to go as far. No matter the change, someone in baseball is going to complain about it. Hitters hate the dead ball. Pitchers hate the rabbit ball. So why is there no simple, neutral ball? For that, we have to look even further back, at how the ball is made.

Currently, Rawlings makes the balls in Costa Rica, about an hour east of the capital of San Jose. In between volcanos and mountains, the city of Turrialba has become something of a manufacturing center, with a population just under seventy thousand. Just south of the main part of town, off Calle 8, one of the larger buildings in town is painted a familiar white, with what appear to be stitches and a clear "Rawlings" sign. Don't expect a factory tour, however. Rawlings hasn't turned this into a tourist trap (they don't even sell t-shirts). While there are two ballfields adjacent to the stadium, only the locals have played there.

Since 1987 when Rawlings left Haiti after an earthquake and political unrest, this is where all the baseballs have been made. The workers come and go, like they do from other factories and shops around the town. There's a fence that keeps people away, and guards, but the gate is usually open, with regular people walking in, walking out, and not looking like pawns in some grand conspiracy.

These are just workers taking parts, putting them together, and coming out with a product that is boxed up and sent away, their time in exchange for a paycheck. Rawlings didn't put the factory in Costa Rica to hide it away. They did it because Costa Rica was cheap and as stable as it comes in Central America. They pay a decent wage and get a decent product. As yet, there are no robot umps and no robot ball-makers either.

While some factories can turn out hundreds or millions of the same thing—widgets, Chevy Malibus, or iPads—baseball doesn't have this. They can get close, but none of the basic equipment of the game can be consistently produced—not the baseball, not the bats, and certainly not the players themselves. How much that variation occurs and affects the game became the object of one scientist who went further down the rabbit hole than most.

Dr. Meredith Wills is in the Hall of Fame because of yarn, but with a skill set that includes a doctorate in astrophysics and a love of baseball. That wild card of knitting was what put her in a place to be the one to unravel (pun intended) the mysteries of the modern ball. Her work revealed that the ball is so much more than horse-hide and yarn stitched together. Instead, it's the inconsistencies in a ball manufactured in mass quantities that may have led to major changes in the game itself.

As people around baseball started discussing the idea of a "hot ball"—one that was constructed in a way to amplify home runs—many around the game were trying to find another reason, from exit velocity to performance-enhancing drugs, as to why players seemed to be hitting the ball farther. Even noted baseball physicist Alan Nathan was taking notice. He gave a presentation at the Saber Seminar in the summer of 2017 discussing changes to the aerodynamic drag of the ball. Seated in the audience, Wills had an *aha!* moment. Wills figured out that her skill set was perfect for taking apart a ball and analyzing the construction, including the yarn.

"In the past, physical changes to the ball have produced changes in performance," Wills told me. "Therefore, it seemed possible that the increase in home runs was due to the ball itself. That was bolstered by the fact that home runs were up across the league, suggesting a source that affected all players. Since I was already good at disassembling baseballs, I just tracked down baseballs from before and during the home-run surge, and I started taking data as I took those balls apart."

Once she did, Wills quickly found significant year-over-year changes to the ball, specifically in the construction of the laces.

Let's address right up front that Wills isn't suggesting some grand conspiracy or that MLB really had much of a goal in mind, at least at the start. "Most of the performance changes I've found are best explained as unintended consequences of economic decisions," she said. "In the one instance where MLB did make intentional changes, the resulting performance appeared to be the opposite of what they expected, and what they told teams."

One of the issues is that there's a huge lead time for the sheer number of baseballs created—1.2 million for the major leagues alone, plus those of the minor leagues and other balls—which makes it difficult for Rawlings and MLB to change things when something is noticed. It can be as much as a year, as this simply isn't a "stop the presses" kind of operation, so a problem now is going to stay a problem for a while and the changes made are difficult to test in quantity.

Researchers on the outside and MLB both figured out that in 2019, the ball had less drag. As Dr. Wills describes it, "MLB officials were aware that the ball had less drag by at least the first week of the season. However, foreknowledge is not the same as tailoring. While leagues like the KBO and NPB have successfully made predictable changes to their baseballs, there is no evidence that MLB has managed to do the same, and deliberate season-to-season tailoring seems beyond the scope of their current manufacturing and testing."

I won't put words in Dr. Wills's mouth, but this is as much about incompetence or at the very least variance as it is about some grand conspiracy to put their finger on the scale for hitters, or pitchers, given what they want to do in a particular season. One ex-player suggested the ball was tailored to hurt the upcoming free agent class, whether that was pitcher or hitter heavy. I'll leave the player anonymous because that's an unsupported idea.

That variation was a major part of why 2017 saw a home-run surge and again when 2021 became a "year of the pitcher," filled

Leather, yarn, and "pill." (Courtesy Dr. Meredith Wills)

with more strikeouts and no-hitters than normally seen, at least through the first couple months of the season. Baseball had made a rules change, or at least an increased enforcement, at the same time, which makes it difficult to separate the two, I believe intentionally. (I'll address that sticky change in a later chapter, but it's important to say here that it was significant.) When MLB responded by saying the balls were within specifications, that really wasn't an answer. According to Wills, it amounted to a loophole.

"Strictly speaking, game balls are 'within spec.' However, that's a tautology, since baseballs that fall outside the official tolerances for size and weight violate the rules. In short, *every* ball is required to be 'within spec,'" she explained. "The implication that 'within spec' is tantamount to sameness and consistency is a canard. Case in point: MLB intentionally changed the manufacturing parameters of the ball for a four-month period of 2020 production, without informing anyone. However, these balls still fell within official tolerances. Had my research not found changes to those internal specifications, MLB would have been in a position to claim that, despite altered performance, the ball was 'within spec,' absolving them of any obligation to acknowledge they had secretly changed the ball."

To know the ball had changed, one would have to look inside and there was very little reason for MLB to do so, given they own the manufacturer, but they also had no reason to let someone like Wills do it on their own. She worked sources, collected foul balls, and crowdsourced the collection of balls for this experiment. Her results, published in a series of articles, first in *The Athletic* and then at *Sports Illustrated*, showed her work, explaining just how MLB had changed the balls, mostly unintentionally. It appears to be much more a manufacturing incompetence than a vast conspiracy.

That raises the question of whether Rawlings can even make a consistent ball that wouldn't have these issues. Wills isn't sure. "At present, that's not an answerable question, at least not for MLB. One thing they have yet to address is the extent to which variation can be minimized. The original Home Run Committee asked for more stringent ball-to-ball consistency, but in an apparently theoretical context. Manufacturing tolerances and limitations were largely unaddressed in their report, and it is unclear whether their recommendations were physically reasonable, or even possible," she told me.

The problem is that doing one thing likely changes something else. "It can be difficult to change one aspect of the ball without affecting others," Wills explained. "While the changes they implemented worked as intended under controlled lab conditions, they altered other aspects of the ball in ways that largely if not entirely counteracted those changes in real-world situations."

That manufacturing process is the issue. Wills looked into that as well, though she, like everyone else, couldn't get inside the Costa Rican factory to check.

"As far as manufacturing, the centers—everything inside the covers—are the most consistent. That includes the pill and each of the four wound layers. This makes sense, since the pills are produced using an injection mold, and the winding process is automated, with a weight-dependent shutoff mechanism. The only human interaction is quality control, where a worker double-checks the weight of each layer," Wills told me. "Baseballs need to be made in a way that's consistent. This is why 2019, the 2019 postseason, and that 1/3 of 2020 being a problem, but consistent isn't the same as identical. If somebody wants a perfectly predictable baseball, they'll be better off playing MLB: The Show instead of watching actual ballgames."

So where's the variance? "The component that shows the largest quantifiable variation is the leather covers. Unlike the centers, which vary by only a few percent, cover weights range from less than 14 grams to more than 19 grams. Considering the allowed weight tolerance is roughly 7 grams, a ball that falls outside of those tolerances is likely to have a cover that is extremely heavy or extremely light."

There was a theory once that after a political upheaval in Haiti in 1986, where the Duvalier family was finally pushed out of rule, the people—and most importantly, the factory workers—felt better and worked harder, perhaps stitching the ball a little tighter. It was known as the "Happy Haitian" theory and it doesn't hold a lot of water, especially considering that we were likely seeing the early effects of steroid use in baseball rather than any sort of political residue.

"You do see variation in stitch tightness, but I'm still working on quantifying that. However, there are historical data going back more than one hundred years showing that stitch tightness impacts performance," she said. "The fact that stitch tightness depends on the individual sewer suggests there will be variation, but there are also several steps of quality control, which should weed out balls with loose stitches."

Dr. Wills is in a unique position, proving time and again that the science is against the narrative that the Commissioner has been trying to sell, not just once, but several times. That puts her in an uncomfortable position, as the whistleblower. At the very least, the scientist born on the day that Hank Aaron tied Babe Ruth for the career home-run mark has hit a few of her own on this front.

You might be asking whether or not a neutral ball even matters. If MLB, Rawlings, or some combination of the two decided this was going to be a year for the hitters or for the pitchers, how much difference could it really make, assuming they could do it consistently?

Wills explains just how big a difference this could create, if it occurred: "There are two ways that a ball can be juiced or deadened. One relates to the 'bounciness' of the ball—the technical term is 'Coefficient of Restitution,' or COR—while the other is associated with drag. Basically, how quickly the ball slows down when it travels through the air. I asked a fellow data scientist, Rob Arthur, about this," she explained. "He confirmed that if the juicing or deadening is due to bounciness, the distance a fly ball travels can vary by a good 50 feet. If it's because of drag, the variation is closer to 30 feet. As for how those two effects combine, we're still trying to find the answer, since changing the bounciness will also change the drag, but not in a way we've been able to predict."

In baseball, 80 feet is a *lot* and even half of that is game-altering. A ball hit 300 feet, plus or minus 40 feet, down the left-field line might (or might not) go out of 20 ballparks. At just half the possible variance, we get anything from home runs to lazy pop flies that the outfielder would likely have to run in on.

The ball matters, and we haven't even addressed how the construction of the ball might alter pitches yet.

If baseball can't make a neutral ball—or if it does, if it can't keep it from being so inconsistent as to not really be neutral—it's not like they'll take that ball and go home. Again, MLB essentially owns Rawlings, and controlling the ball, rather than simply making the rules for what the ball is supposed to be, is one way that they could maintain control of the game itself, even if they're not good at actually making the balls.

In one sense, they are. The balls look fine and are made in quantity. They likely make some profit, though Rawlings doesn't have to share figures as a privately owned company. If I tossed you one of these balls, one of the ones from Meredith's secret stash that she hadn't cut open just yet, it would be fine. We could have a catch, we could hit some, and the grounders would roll true. If Mike Trout decided to sign it, that'd be fine too and no one would question it.

But that's not the issue for MLB right now. It's a question of how the ball they use affects the game they play, and how that affects who watches, who buys tickets, and to some level, perhaps even how well players do in categories that will get them big new contracts. Yes, that's as much of a conflict of interest as it sounds.

Top that with some egregious statements from Commissioner Rob Manfred as the ball issues were called out, and while it's again not conspiratorial, it's hardly benign either. The issues of the very fairness of the ball are suddenly in question, where MLB is the arbiter and the constructor. Pitchers don't call strikes themselves for a reason.

Let's take this one step further and be careful, as it's a slippery slope. Imagine someone slips the workers in Costa Rica some dollars to sneak out a sample of 2022 baseballs. This nefarious soul gets his hacksaw and tools, discovering quickly that next year's ball is a little bit better for hitters. He flies to Vegas, hits the over for the Yankees, who have more sluggers than most teams and could see a bigger jump than most, and makes a couple more bets on home-run totals. This doesn't take an Arnold Rothstein or a conspiracy to see how baseball construction and betting totals, a growing part of MLB's business, could come into conflict.

The trust that MLB holds or loses is an absolute key. Losing public trust has been at the heart of most of baseball's crises over the years. If Jim Bouton's final words of *Ball Four*—"You see, you spend a good piece of your life gripping a baseball, and in the end it turns out that it was the other way around all the time"—turns out to be false, that we didn't really know what any of baseball's players were gripping, then there's a chance baseball has its biggest crisis of confidence since the Black Sox. If the baseball itself isn't consistent if not neutral, what then is the game of baseball?

2
THE BAT

The bat is not quite as fundamental as the ball, but it's as basic and as iconic. From Wonderboy in *The Natural* to something as wildly different as the gangs in *The Warriors*, the baseball bat is known worldwide, even in those countries that don't really play baseball. If there's a platonic ideal of the bat, it's likely a turned piece of ash, fired to make the grain stand out and then branded "Hillerich & Bradsby."

The makers of the Louisville Slugger have been turning out bats in Louisville, Kentucky, for over a century and they still do it there, though the process has changed a lot. Other manufacturers have gotten in the game, but you can still go to downtown Louisville, walk in under the giant bat outside the door, and see the process of making a major-league bat that isn't too far off what they did for Pete Browning and Honus Wagner.

As with the ball, the rules about a bat aren't specific. There's certainly more to them than the ball, with this being the key portion:

"The bat shall be a smooth, round stick not more than 2.61 inches in diameter at the thickest part and not more than 42 inches in length. The bat shall be one piece of solid wood."

There are four parts to Rule 3.02, but they can be very simply distilled. The bat must be one piece of wood of a certain minimum and maximum diameter. It can't be colored, unless approved, and it can't have a cup at the end of more than an inch. That's really about it, which leaves a lot to the imagination.

There's nothing that says what kind of wood. There's no discussion of length, shape, or weight. "Solid" does rule out some of the chicanery of altering a bat, such as "corking," where the bat is hollowed out and refilled with another substance, usually cork but sometimes rubber, to re-weight it or make the bat have more "bounce." Most of it didn't work. Apparently, snake oil is a common ingredient in a lot of bats.

What's happened to bats is mostly that they've gotten smaller. Babe Ruth reportedly swung a 48-ounce bat, and that's simply unthinkable to most modern hitters. (According to Jon Shestakovsky of the Baseball Hall of Fame, one of the Babe's bats that they recently showed off to MLB Network weighed "only" 38 ounces!) Cap Anson, the White Sox captain in the 1890s, swung a "manly three-pound bat" according to Gary Gillette. Heinie Groh famously tapered his bats, leading people to call it a "bottle bat," but as this picture of Ollie Pickering from 1907 shows, those bottles really look more like table legs. You'd barely recognize it if you were used to modern equipment.

(Courtesy John Thorn and his Our Game blog)

While there have been hundreds, maybe thousands of innovations in the construction of bats, largely in terms of metal and composite bats used at lower levels, there have been very few innovations that have caught on in the major leagues. In fact, there are really only two major ones.

In the 1990s, a shift from ash to other substances started to happen. Some of this was based on the need to diversify due to an invasive, destructive beetle called the ash borer which threatened the supply, especially in the key Pennsylvania forests that had been used since the days of the original Louisville Sluggers. Harder maple and other wood types could be fashioned by more modern tools into bats that weighed the same and qualified under baseball's admittedly vague rules, but had other properties such as hardness or density.

There was a secondary move to try and create so-called engineered woods. These are common wood types and regularly used in construction. Constructing long and strong pieces of wood for headers and joists led to glue lamination techniques. These "glue-lam" boards can be as strong as steel beams and cheaper, put together with layers upon layers of wood and glue, then cured. The same is true in floors, where engineered woods have largely replaced hard woods due to the ability to both customize the textures and produce more from less, an issue with both cost and environmental concerns at the forefront.

The second innovation is to the handle of the bat. Axe Bat took a literal axe handle and applied it to a bat. The knob of the bat changes the hand and wrist geometry, which gives a different feel. Some like it and some don't, but the major issue is that it purportedly protects the wrist from biomechanical stress. There's some controversy there, but for some, it really is a positive change and the feel alone is enough to make some evangelize the bats. The patented technique can't be copied, but there haven't been any other major changes where makers try something different to get around the patent or just improve altogether.

The next innovation in bats is happening now. Using technology and knowhow, the cutting edge of hitting starts with customized lumber. Almost all major-league bats are made to precise specifications, but the differences that exist haven't been able to be noted, sorted, or even exploited. That's changed now.

LongBall Labs is a great, great name. Keenan Long should just thank his parents for setting up his career from day one. Explaining exactly what that is makes for a bit more of a challenge. LongBall Labs is more than just a "bat fitter." Those exist, helping to fit a bat to a hitter in much the same way that golf clubs can be customized to a golfer. It's simple to help someone figure out the proper length and weight, how the grip is set up, and the like. What LongBall Labs does is something far more important.

Most batters and bat makers have looked at weight and length, and others focus on feel. Some of that, like the Axe bat, is innovative but most of these qualities are aspects like diameter and how the grip feels, which can't be scientifically based. There have been some attempts, with measurements of "comfortable grip strength" that have largely been discarded due to a lack of results.

Long and his company look beyond that and into the very fiber of the bat. Yes, fiber. Wood is a fibrous, porous plant product that comes from certain trees. Saying that it is a "composite of cellulose fibers" that are encased in a "matrix of lignin" to resist compression already sounds like an engineer has been involved. However, every tree is not the same. Every piece of lumber that comes from those different trees is not the same. A wood bat is not a wood bat is not a wood bat, and that's where Long comes in.

"Length, weight, turning model, and wood species are the four main variables in a professional wood bat," Long explained to me during a long conversation. "The turning model has a few subcategories within it: knob geometry, taper section, barrel length, and whether or not the player wants the bat to be cupped. Aside from that, everything else is constrained by the rules. Also, the four categories also constrain each other. For example, you could specify the length and turning model, but then you might have trouble also specifying the weight of the bat because you had already constrained the volume by designating the length and turning model. From that point, your only chance to change the weight would be to adjust the species of wood or hope for a particularly light or heavy lot of billets in that particular wood species."

The key here is that while there's rules and constraints, there's also no push to a single format. "Baseball has learned that the optimum qualities for a bat are not absolute qualities. There's no science-based approach that has ever proclaimed 'this one type of bat is best for all players' or even 'this one particular feature of the bat is best for all players.'"

Worse, the variance is huge. In a May 2021 article on LongBall Labs in *The Athletic*, Eno Sarris quoted an anonymous player: "One major-league hitter said off the record that he'd gotten bats that were labeled wrong in inches, labeled wrong in weight, stamped wrong, and the only thing they got right was the color."

Just as there's no one bat to rule them all, there also hasn't been much guidance for the players. Technology is changing that, as bat speed is now not just a guess. Sensors like Blast or camera-based systems like HitTrax can get very specific bat speeds and trajectories, even measuring over a session how much fatigue factors in. Until recently, these were done on feel and to some level, results. The hits would tell you which bat worked, but never predictively.

Long explained how those come together with bat selection. "Typically, a hitter is left to their own devices to optimize for two outcome-based measurements: swing speed and exit velocity," he said. "In layman's terms, your absolute best bat will always end up being some tradeoff between the bat you can swing the fastest and the bat that produces the highest exit speed for you. The hitter must take into account both of those outcome-based measurements."

The rough equivalent to this would be to have Jon Rahm or Bryson Dechambeau reach into their bag blindly, maybe pulling a state-of-the-art driver out, or maybe a persimmon wood like Bobby Jones or Arnold Palmer used. Worse for baseball players, the ball isn't on a tee for their swings, good or bad wood aside.

Finding that bat has been trial and error, at best, for batters and the threat of snapping one at any given time makes optimization difficult except in the most basic of parameters. If this size and weight feels good and it snaps, give me another just like it. Since bats tend to come in dozens, the theory is that any bat from a single batch should be roughly identical, especially with modern quality control and manufacturing techniques. This isn't Ben from HGTV making a table leg in the Scotsman shop, as nice as those are.

Just the difference between ash and maple can lead to interesting differences. Long explained to me that ash is an "open grain wood" while maple is "closed grain." What that means for a hitter, even a specific hitter, can be a lot. I remembered always being told when I was younger to hit with the brand of the bat up. (The brand is usually the famous Hillerich & Bradsby mark, which is a literal branding, burned in on their wood bats.) For ash, that is key. "A hitter wants to have the brand up and hit the ball between the brand and the bottom. For open grain wood, hitting it with the grain is key, otherwise the wood is going to fail. It will flake and be much less durable."

For maple, it doesn't matter. "Delmon Young was a monster back in the day," Long said. "He would go up to bat and he was so big and strong, but he had a lot of quirks at the plate. He'd shuffle and lick his lips, but he'd twirl the bat in his hand, spinning it around. People would say no, no, that's going to ruin the bat, but he was using maple. It's closed pore, so it doesn't matter. He could hit with any part of the bat."

Beyond feel, weight, length, and wood, finding that right bat becomes guesswork for most. Trial and error can work, but it's guesswork. Is a batter swinging the best possible bat in the biggest situation? Not if he gets the balance wrong. Long says it's a big difference. "If you only go for the bat that results in the highest swing speed, you would end up swinging a 'broom' that would do very little damage when you made contact. It would hit the ball weakly, producing very low exit velocities and would be likely to break often. If you only go for the bat that produces the highest exit velocity, then you most certainly will end up reducing your contact rate because you'll be swinging a 'sledgehammer' that would most certainly be detrimental to your in-game performance. Where the absolute best bat for each player lies is currently up to the player themselves to find. There are some early-stage technologies being used to help the hitter optimize."

Brooms and sledgehammers have their uses, but not as bats.

The desired outcome is also a consideration. There aren't many situations where a batter should go up swinging for the fences, but what about something like the Home Run Derby? Could someone challenge Pete Alonso by having a better bat? Long thinks so. "It is definitely an optimization problem. If we are talking about hitting a batting practice fastball in a home-run derby situation, then the bat with the heavier end loading will likely outperform the other bat, assuming that the difference between the two is not too massive."

Again, the Home Run Derby doesn't translate to games and Long acknowledges that as well. "In games, however, it could easily be argued that the lighter swinging bat would be better considering the ever-increasing whiff rates and strikeout rates in Major League Baseball. A bat that swings faster for a given player allows that player more time to view the pitch and offers them a chance to commit to their swing later. The later a hitter can commit to a pitch and make contact, the better hitter they will be."

Long uses proprietary scientific techniques that help him show batters which are "better" wood than others. As he puts it,

Keenan Long (far left) and the crew from LongBall Labs. (Courtesy LongBall Labs)

"Everything we've done is built around weaponizing the ability to select for that one discrepancy—wood quality."

Long uses a measure he calls RIPz, which gives a score to each bat based on proprietary factors. I got him to explain a bit more. "Players have known for years that a shipment of identically labeled wood bats were not actually identical. Just by looking at the wood itself, it is apparent that the bats are different. Players have tried many different methods to find their 'gamers' amongst the duds in their bat shipments," he shared.

"LongBall Labs was the first to quantify what these bat differences mean for performance and turn that insight into a performance advantage for MLB hitters. The original discovery that started it all was that volumetrically identical bats (from the same shipment of identically labeled bats) exhibited significant performance variability," he continues. "Basically, bats that looked and felt the same to the player were performing very differently.

To quantify this discrepancy and give the player the best chance at picking up their optimal bat, LongBall Labs created a new metric, RIPz." It's pronounced "rips." "Every single bat in a shipment gets its own RIPz score which tells a player how that bat would perform in their hands. Armed with this information, hitters could test various bats to find their ideal RIPz score—the bat that produces the highest exit velocity for their swing."

In essence, RIPz is a metric that quantifies the minute variations in the physical properties of the wood so that every hitter can optimize their performance. RIPz and LongBall allow the batter to get what they paid for, using the best bats, allowing both more exit velocity and more consistency.

Scientific bat selection is just the start of this. It's easy to imagine a manufacturer licensing RIPz, sending out only the good bats, and turning the rest into pellets for my barbecue smoker.

Any baseball player is going to think about what can be done to make a bat better. There are tons of unscientific techniques, from "boning a bat," where the player basically rubs a chicken bone all over the bat, part sanding and part heathen ritual, to something illegal like corking, where the core of the bat is taken out and filled with cork to alter the weight and density. Long says none of them work.

"Even if corking a bat were suddenly deemed legal for all players, I still would not recommend it to any of my clients. The idea that any type of springy material would have an effect on the collision is one of baseball's enduring fallacies," he says with a laugh. "The quickest and easiest way to debunk the fallacy that putting a springy insert into the core of a bat will increase performance is to know one thing about the collision of bat and ball in the big leagues: the collision only lasts about one millisecond. What this means from a physics perspective is that the ball never 'sees' the reaction from the springy material because it is already off the bat before the impulse from the initial contact even makes it to the core of the bat." High-speed video backs this up.

Long thinks it goes deeper than this basic physics, however. "This is the same exact reason why grip style, at the point of contact, has absolutely no effect on the batted ball outcome. Whether you hold the bat tight or completely let go of it before contact, it simply does not matter because the bat's reaction to those things does not occur until after the ball is already off the bat," he showed me. "Going back to the corked bat concept, the ball can't 'see' the springiness of the cork, but it can 'see' and does react to the difference in weight created by the corking. Removing the wood and replacing it with cork lightens the overall weight of the bat. That is something that does indeed affect the collision with the ball." But not for the better!

Boning a bat is the same, though Long says there's a bit more effect of this. "[Boning] increases the surface hardness of the bat, but does not have a discernible effect on exit velocity. Boning, along with the recent surface finishes that have been developed by high-end wood bat brands, serve to increase the durability of the bats, but again, not the performance." These bats can be tested with modern sensors and cameras to measure what we really want: results. "We now know, however, that exit velocity can be accurately measured and used to compare the performance differences between given bats or between given players."

Then again, one online forum suggests that because chicken bones are hard to come by (they are?) that an equivalent way to make it work would be to "use a piece of porcelain or a sturdy bathroom appliance." Toileting a bat doesn't sound like something anyone wants to try.

Long believes that the biggest change could be specialized bats. "In the near term, we will see players make in-game bat adjustments. The old concept of situational hitting will soon apply to the player's bat choice. For some of my clients, it already does! I make this prediction not based off what I think will happen, but based off what both young and old big leaguers have told me in video calls. I was surprised the first time I heard it and thought it was a fluke, but this same thing was brought up without prompting on several occasions. I think in the next couple of years, the players will know the exact exit velocity potential of each individual bat in their bag. Thats what we do at LongBall Labs."

According to calculations done by MLB's Tom Tango, a longtime top sabermetrician, that added velocity given by selecting the right bat could be the difference between being Jeff Bagwell and being Mike Trout. That may not seem like a lot, but just look at their lifetime stats. The difference for one hit might be between the warning track and the seats; for a career it might well be the difference between Cooperstown and not.

But what about aluminum, metal, and the newer composite bats? Frankly, they're never coming to the major leagues because even with modifications, the bats are still "hotter" than wood. Recent regulations at the collegiate and high school levels have made the bats play truer to wood, but at the same time, the original advantage of metal bats has broken down. Metal and composite bats break, get out of alignment, and develop weak spots almost as quickly as a wood bat, all at a cost that can get into four figures.

The original idea was that a cheap metal bat would cost less and for a Little League team, a duffel bag full of different sizes would be good enough for a couple years. That's just not the case anymore as the market has shifted from "good enough" bats to the colorful, powerful bats of today.

Where there is an ability to make a bat "better" with metal or composite, the rules have constrained them. Starting in 1998, a standard called BESR (balk exit speed ratio) was the first to address the issue of how fast the ball was coming off a bat. Using a test and some math, a bat needed to be under the magic number in terms of ball speed coming in, bat speed, and exit velocity. The problem with BESR is that it was only done on one test pitch and manufacturers quickly learned how to manipulate it. Composite bats often have a "working in" period where they actually get harder and bouncier. There actually exists a black market for some bats built in the late 2000s that look more like the price of a nice car, despite the fact that they're illegal for use in most games.

BESR was replaced in 2011 by a test called BBCOR (baseball bat coefficient of restitution). It's less a test than a measurement, where coefficient of restitution—which we discussed with balls earlier—is determined. The test takes a stationary bat, held at the grip end, and fires a baseball at it at 136 mph. The COR cannot exceed 0.50, which means, roughly, that the ball can't come off at more than 68 mph. The calculation is vastly more difficult and adjusts for a lot of things, but bats are tested to not be trampolines.

With these limitations in place and costs for standard metal bats going up and up, wood bats are becoming more of a consideration. In the fall of 2021, the top-rated bats by *Bat Digest* were the DeMarini "The Goods" ($399) and the Louisville Slugger "Meta" ($499). Most of the other top-rated bats ranged from $229 to $599 and none came with a warranty of more than one season. That's a long way from that Hillerich lumber Pete Browning swung.

The costs and the profit are why the top companies continue to refine the bats, but at best, they're limited by the rules in place. If MLB isn't going to switch and Mike Trout isn't swinging the latest composite monster, it's a tougher marketing scheme. Which is one reason things are starting to shift the other way.

With changes to the minor leagues and to offseason (summer and winter) baseball, there is more of a shift to wood already. New summer leagues like the Draft League have joined other leagues in transitioning strictly to wood bats. Again, the cost is not significantly different even over the course of a season and scouts love it. For players already paying in the thousands for the opportunity to play, a box of a dozen bats is just another expense they'll have to bear.

With new techniques, there's the chance that there's going to be a middle ground. Some sort of engineered wood bat may end up having the best properties of both. The engineering necessary exists, but it is outside the rules of baseball and there's been little reason for them to change and adopt any sort of innovative technology. Baseball is a traditionalist game and the sound of a ball on a wooden bat is as iconic as any. As long as there's wood and there's baseball, MLB bats will be wood, but as I've shown, science isn't going to stop just because the bat is the same wood that the Babe swung.

3

HITTING

"It is high. It is deep. It is gone. Home run!"

That is John Sterling's home-run call, virtually every time a Yankee has hit a home run for the past thirty-something years and five thousand games. Add in a little bit of his colorful puns on people's names, a true love it or hate it thing, and his call is not only famous, but accurate. Just not precise.

Until 2005, teams guessed or at best had very imprecise estimates for home-run distances. One Manny Ramirez home run was estimated at near 500 feet, and Greg Rybarczyk knew that was wrong. Over the next few weeks, he worked on a system that allowed him to not only tell that Ramirez's shot went "only" 449 feet, but had an accurate system for every other park. ESPN came calling pretty soon and Rybarczyk's Home Run Tracker became the gold standard for years.

Today, those homers are called not by guesses or even an elaborate calculation, but measure to the inch by cameras and radar models of the stadiums. They're adjusted by barometric pressure, the effects of any wind, and a number of other variables, all just seconds after the ball lands.

In fact, the ability to track things in real time is starting to become a time machine. Teams have been working on systems that are descended from complex models that help predict weather in order to try and predict hits. If a certain pitch is thrown to a certain hitter in certain conditions, there would seem to be nearly infinite possibilities for what can happen, but one MLB analyst (who was not authorized to talk about his work for a team, so he could only speak if granted anonymity) says that's not really the case. "There's a controlled number of variables," he explained. "We know where everyone is and where the batter hits that kind of pitch, if we have enough of a data set. If we don't, we just shift it to the type of hitter and the data set gets bigger. Fielders can only move so much. Pitches mostly do what they usually do. Hitters have tendencies."

At that stage, it's not a guess but a question of probabilities, and a bigger question of computing power. "We've been doing this about a decade, but we've only recently gotten to a point where it's forward-looking. Previously, we were trying to predict patterns and fielding positions. Now, those are all known and can be calculated."

The biggest shift, he told me, is not in the calculations, which have become pretty standard for his team. Instead, it's in the ability to get the information out quicker. "We tried an app [phone application] a couple years back, even knowing that [MLB rules] don't allow it. Beyond that, how do you transmit that information and make it readable, understandable, and most importantly, actionable?"

The team went from their app to a system of lights on a small box. It's better but also against the rules of baseball to transmit things down in real time. "At least it is now," he said. "If things change, we're ready. The biggest issue we've had with the testing is not trying to be a better trashcan as one of my bosses put it, but being useful. We're still not there. This system's one of those back burner things we work on here and there, but while we've used it with some minor leaguers, I don't think most of our major-league guys have any idea. We'd never use it up here until it was rock solid and certainly never unless rules change."

So while some are looking forward, others are looking virtual. It's been relatively easy to track the baseball for the last fifteen years, either with radar systems like TrackMan or camera systems like Hawk-Eye. Rapsodo, best known for its pitching system, also has a hitting system that can track things like exit velocity and launch angle. But the real revolution is taking place on a smaller scale, inside, and while still expensive and specialized, the ability of a couple technologies to take baseball to new places is really giving hitting a chance to have an advantage pitching will never have.

Let me explain that last statement further. For years, there has been a shifting balance in the war of pitching versus hitting. Some years, the pitchers are on the high road and other years, the hitters are clearly getting the best of them. There have been varying explanations, from the ball and bat, as you saw in previous chapters, or just an influx of talent, a "class of pitchers," or even something like expansion diluting the pool of both.

However in training, hitters always have one advantage. They don't get tired as fast. Pitcher workload is always an acknowledged issue, but there's no similar estimate for hitters. Nevertheless, hitters do fatigue, even in games where swings are minimal but they're often standing in the hot sun, running, and just waiting in game. The batting cage is an area where fatigue can be almost discounted and most sessions are set up in rotations, taking eight or 10 swings then letting a teammate step in.

Add in a pitching machine and a batter can, in theory, go on forever. But hitting in a cage can get boring and monotonous, nor does it replicate the real experience. A hitter can see pitches, feel the ball come off the bat, but what if the process could not just be enhanced, but become a game?

I hate when video games and the like count the hours you've played, because I start thinking about all the things I should be doing rather than playing Forza Horizon or Clash of Clans. However, when players sit there for an hour, hitting in the cage with a HitTrax machine, I don't think any of them regret it.

I'll date myself here, but the HitTrax unit looks like one of the once ubiquitous *USA TODAY* newspaper machines. It's a microwave oven–sized box on a smaller stand. One side has all the controls, plus a connection to a computer and screen, while the other contains a couple cameras that can track the ball, coming in and going out.

Standing just outside the cage and watching a hitter use a HitTrax adds a new dimension to it. The rhythm changes, slightly. The hitter stands in, watches the pitch, hits it, and quickly turns to view the large screen set up just outside the cage. He's picked a park—say, San Diego's Petco Park or Fenway in Boston so he can see whether he can clear the Green Monster. The hit is immediately evaluated and in real time, the ball goes where the ball would have gone in real life. The screen shows the ball that the player just hit rolling to second, or sailing over the wall depending on launch angle, exit velocity, and more, all in real time.

I was able to watch HitTrax in use with University of Indianapolis player Caleb Vaughn. The young power hitter has had over one hundred sessions on the HitTrax and he called it "addicting" (as he laughed). While he hit, we talked about what he loved and why he found it so compelling. "It is so realistic," he said as he hit a long fly ball to left field. "I get instant feedback of how hard, far, and the spot that you hit the ball. I just want to continue using it until I can't anymore."

And he does, as a worn pair of batting gloves attests.

I asked Vaughn if the machine was tricking him into making him work harder. He paused and nodded. "Baseball is such a mental game that could easily be the case. If you're feeling good, smacking baseballs on HitTrax, and the feedback is as accurate as it is, that can only bring confidence. Being confident and seeing what you can translate from the system onto the field is a beautiful thing. It's like 'if I can do it on this machine, I can go out in a real game and do the same.' I am proof of that." If you need more proof, there's a video on YouTube of Vaughn hitting a long home run in the conference tournament that might convince you.

Vaughn continued to swing—and look at the screen with every swing. He stopped occasionally to tell me what he thinks of HitTrax. "One thing I will say is the fact that HitTrax is so much fun and can be used at any time. It makes you want to go up to the cages and get some work in. It makes you want to go grind and get better. It's an attraction. The fact is, if you put in the work, you're bound to get better and HitTrax makes it fun to work your tail off. Having fun doing what you love—what more could you ask for?" (I checked, and Vaughn is not an employee of the company, but I'm sure they'll be glad for the endorsement.)

While many colleges, teams, and facilities have bought the units, which run about $20,000, many others will pay to use them. Some facilities will essentially rent out their HitTrax unit, letting other players come in and use it, albeit at a cost. Helping to offset that initial outlay makes it a bit more accessible, but there are competitors out there trying to make a similar unit on the cheap. I haven't seen a good one yet.

While few would argue that using something that makes a monotonous and repetitive task into a game that players are willing to pay to play is a bad thing, the type of training that HitTrax allows—tuning the launch angle to hit more homers—is somehow controversial.

Some arguments you just can't win though. In the hitting world, the latest old versus new argument centered around launch angle. Of course, hits have had a launch angle forever—this simply means that a ball leaves the bat at a certain angle, a certain trajectory. Studies have shown that a 29-degree launch angle is the most common for home runs in the major leagues.

This is no different than the lofts on golf clubs. Certain clubs hit at certain angles, which create different outcomes. Add in spin and all the other effects that happen to a ball and a hitter can, in theory, optimize for outcome by tailoring his swing in the same way that a golf club is tailored for a certain outcome. It's not as predictable (no tee), but there is an element of advantage that can be gained from training for a certain launch angle and creating the probability that hits are closer to that ideal more often.

To take the golf analogy a bit further, a 29-degree loft is what you'd find somewhere between a 5 and 6 iron. A 5 iron is typically around 27 and a 6 iron around 31, but there's some variation from manufacturer to manufacturer and fittings for modern players, especially pros with huge club speeds and distances, can change those. Let's say 5 iron for ease, since there's not a 5 1/2 iron handy.

A 5 iron is normally used, for a low handicap golfer, in a shot between 170 and 190 yards. That's 510 to 570 feet, which is longer than almost all recorded homers, even in a home-run derby. Of course, the ball is smaller, harder, and more aerodynamically suited for distance. Even with the swing being different—a more arcing underhand motion for getting the loft on a golf ball versus the violent, near linear swing of a bat—this is remarkably similar in terms of distance, given intent.

As important as launch angle is bat speed, which as we discussed in the last chapter, is tunable and trainable as well. It's more than simply strength, but it is a tunable skill where a hitter can gain an advantage by optimizing. The combination of bat speed and launch angle, along with the best bat, puts the hitter in a more positive position when contact is made.

In fact, it's required that both of these are done together, because the optimal launch angle is dependent on bat speed. Hit a ball at 20 degrees at 80 mph and it's a soft liner to an infielder. Hit the same pitch at 100 mph and the same 20 degrees and now it's more likely to be a hit.

Flat out, hitting the ball harder is better. Hit it over 100 mph and even a well-placed infielder is going to have almost no time to react, let alone move to get the ball. Beating a shift—and we'll talk about those soon—is best done by hitting the ball harder, not hitting the ball differently.

Interestingly, there's more and more research that says while hitting instructors are attempting to tune their hitters to certain launch angles, the actual launch angle is more controlled or at least more affected by the pitcher. One stat, inelegantly called aLAA

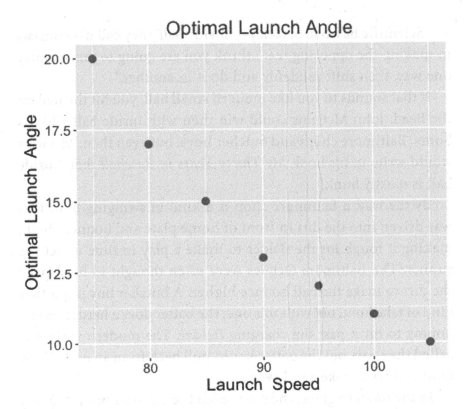

Optimal Launch Angle

(Courtesy Jim Albert)

(average launch angle against), seems to be more predictive than a hitter's average launch angle.

The lesson here is that while people argued and misused the concept of "Launch Angle" as a misguided attempt to discredit modern metrics in hitting, it was always more important to hit the ball hard. This data simply showed it, in black and white. While I say "you can't argue with data," people always try. Tailor your swing and it makes it tougher for them to argue as you trot around the bases after putting one in the cheap seats.

In 1911, the *New York Times* described what they called the gold standard of baseball tactics. Since the 1890s, it had been called "scientific baseball."

"Scientific baseball of today—'inside ball' they call it—consists in making the opposing team think you are going to make a play one way, then shift suddenly and do it in another."

If that sounds to you like modern small ball, you hit the nail on the head. John McGraw could win then with inside ball with its bunts, Baltimore chops and butcher boys, but even then, McGraw would write in his book *My Thirty Years in Baseball* that "inside ball is mostly bunk."

By the way, a Baltimore chop is a kind of swinging bunt that was driven into the dirt in front of home plate and bounces high, making it tough for the fielder to make a play in time to get the runner. The Baltimore Orioles were often thought to hard-pack the dirt to make the ball bounce higher. A butcher boy is another kind of fake bunt, but with this one, the batter does a harder swing, hoping to hit it past the charging fielder. The modern variant is called the slash, but the batter has to pull back from an early show bunt and try to take a full swing.

In the modern game, the bunt, even the sacrifice bunt, is going the way of the dodo. There are whole charts created, called run matrices, that show when bunting to move a runner might make sense. Bunting can be summarized in this way in terms of its use: "almost never, maybe at the end of the game when tied or down by one."

This chart, created by Dan Blewett (tip of the cap), shows the run expectancy of different base states—who's on and what bases. It doesn't take into account the inning, the hitter, the pitcher, or even the runner. This displays broad-brush averages for how many runs a team should score given a certain situation.

A common bunt, sacrificing the runner from first to second, giving up the first out of an inning, shows a loss of runs expected, going from 0.859 to 0.664. Outs are precious, as are runs, and the manager who just waggled in that bunt cost his team both in the broad sense.

RUN EXPECTANCY

BASE RUNNERS				EXPECTED RUNS		
1B	2B	3B		0 OUTS	1 OUT	2 OUTS
-	-	-	➜	0.481	0.254	0.098
1B			➜	0.859	0.509	0.224
	2B		➜	1.10	0.664	0.319
1B	2B		➜	1.437	0.884	0.429
		3B	➜	1.35	0.95	0.353
1B		3B	➜	1.784	1.13	0.478
	2B	3B	➜	1.964	1.376	0.58
1B	2B	3B	➜	2.292	1.541	0.752

Data Taken From Tangotiger.net, MLB Data 2010-2015

An Aside on Percentages:

Anything with percentages means that anything can happen, but some are more likely, even much more likely to happen. That doesn't mean they always do, or that the low-percentage play doesn't work. In fact, when they do, you'll often see people say, "That's why your chart doesn't work," as if there was a Calvinist destiny for Buster Douglas to knock out Mike Tyson. I usually leave that discussion, but when drawn in, I'll remind people that sometimes, you do roll snake eyes, but I wouldn't bet on it. It's smarter to bet red or black than expect to hit 15 on roulette. The model over there might go out with you, but then again, some percentages are in fact zero.

In almost all of baseball, teams and players are simply trying to tip their percentages in their favor. As Crash Davis reminds us in *Bull Durham*, the difference between .250 and .300 in the big leagues is one hit a week. One dying quail. One seeing-eye grounder. Anyone not actively trying to find their dying quail equivalent, in baseball or anywhere, simply isn't trying.

Bunts and steals have fallen by the wayside because of percentages and the fact that even people that don't run the numbers can see that the best way to score runs is to hit the ball over the fence. Baseball's more complex than that, but "ball go far, team go far" is supported by the math (and less by the grammar).

What's a bit more surprising is that the rise of the strikeout is another one of those percentages that had to be put into context. Strikeouts aren't bad, aren't humiliating, and aren't considered a major negative in today's game. It wasn't that long ago that they were, and that coaches would light up a player, or worse, for watching a third strike go by. Today, a strikeout is just another out, and the easier percentage to look at is on-base and slugging. Do enough of those, and no one's going to criticize those strikeouts.

It takes run expectancy matrices, supercomputer calculations, and some old-fashioned baseball experience to make it all work, but there are so many details, so many things that can go wrong, it's no wonder that hitting is really about managing failure. A .300 batting average might not mean as much in light of more advanced stats, but it's 70 percent failure and that holds true from Pete Browning to Babe Ruth to Shohei Ohtani. Dealing with failure is one thing, but managing it is different and that requires a more considered approach, somewhere in between superstition and simple routine.

One of the rituals that's been around in baseball for who knows how long is batting practice. There's a ritual element to it. For the players, it's a way to lock in and part of the rhythm of game day or practice. For the fans, it's fun watching players launch them into the seats before a game. It's almost like watching the trailers before a movie.

But does it work? That's something that's been questioned over the last few years, with more teams at all levels looking at what's best for helping their hitters get ready for a game or find their groove as they slump. Baseball has never been a sport about practice as much as it is about consistency. Changing something as ritualistic as batting practice is a tough sell for players that have grown up on it.

One of the first teams to try something new was the Toronto Blue Jays. Every team has a coach or two that tosses batting practice. Tommy Lasorda did it for years with the Dodgers, famously firing in fastball after fastball ahead of games, even when he was managing. The Jays went away from the soft tossers and hired several former pitchers, some out of their own organization, and then got them to do something very different: impressions.

The Jays' idea, one I'm told came from Assistant General Manager Joe Sheehan who climbed to his current role after shining as a data analyst for the Pittsburgh Pirates, is to get several pitchers, righty and lefty, that can go through a number of arm slots, throw a number of pitches, and essentially imitate, to some extent, the upcoming pitcher. Players watch endless tape on pitchers, looking for clues and tips, to see the shape of the curve and how it breaks on them. They'll watch their own at-bats and recent at-bats against them. The Jays simply wanted to translate that in some way and the results were good, but it's tough to isolate statistically. For what it's worth, they're still doing it.

Other teams have shifted to a more realistic batting practice, if not an imitation. Several squads have taken to using their own pitchers at full speed or, more often, a pitching machine tuned up into the high 90s. Some teams have even more advanced machines in their stadiums—"down in the tunnel"—that can replicate pitches in high fidelity. Called a robot rather than a pitching machine, the Trajekt Sports system pitches way better than it spells.

Everyone has probably seen a pitching machine at this point. The standard Jugs machine, with its two wheels and a "gun barrel" at the front, has been in use for fifty years and does the job. Spin those wheels up and it can fire a ball as fast as you'll ever want. What Trajekt, another innovative baseball company out of Toronto, does is different. In fact, they don't even call their big black box a pitching machine. It's a "pitch replication robot."

That robot, or pitching machine, or whatever you choose to call it, is also really a black box in both senses. The machine itself is a big rectangular box on wheels, a hole in front to shoot the balls out and one in back to put them back in. What happens inside is unknown, but we know the inputs. A Trajekt machine takes the parameters seen on Statcast—MLB's system that shows data like velocity, spin, and more—and replicates it exactly. You don't push a button for fastball here. You push a button for Marcus Stroman's fastball in the third inning of the game on June 17th to Mike Trout.

If you're Mike Trout—or any lesser human hitter—you can immediately see the value. That pitch that got you out, that tied you up with men on first and third, you get to face again and again. A batter can face a perfect sequence, something no mere imitator can do on their best day, and never get tired. Sure, it's coming out of a black box rather than a pitcher, but maybe you need to be able to adjust to Jacob deGrom's slider after his fastball. Click, click, and you can face that, in any sequence he's ever done.

A Trajekt robot doesn't come cheap. You could hire a couple minor leaguers to imitate Stroman, deGrom, and anyone else far more cheaply, but they get tired, they don't replicate perfectly, and they need to go home to sleep. I'm told there's one major-league hitter who has one at home. I imagine him walking downstairs after midnight. The wife and kids are in bed. He opens the door, pushes a couple buttons, and he's facing everyone he'll face next season every night until he becomes their nightmare, mopping up hit after hit.

Even with a pitch replication robot though, batters can never get enough practice. With this in mind, the company with a lot of major-league connections is bringing baseball into the virtual world. WIN Reality looks like a simple (and goofy) VR headset. In fact, they use the standard Oculus Quest unit, sold by Facebook.

Instead of being a Fruit Ninja or light sabering with Darth Vader, WIN Reality has created a system that simulates facing pitchers.

In the virtual world, the pitcher never tires.

WIN Reality has a library of thousands of pitchers and pitches that can be called up, plus training modes that work on a variety of skills. It recently launched an update which uses a small attachment that slides over the bat and allows the Oculus controller to be used with a full-size bat. (For a lot of WIN Reality's market, I'm not sure swinging a full-size bat in the living room is a great idea.) There's a slight weight and balance issue, but people I have spoken with that have used the new system say it has a better feel than just using the controllers, which didn't approximate the size or feel of a real bat.

There is some haptic feedback, but WIN Reality has yet to create a real "ball feel" on a virtual hit. It's still a very accurate simulation and far better than anything else on the market. The iterations the company has gone through over the past few years are impressive, and as the headset and tracking improve, WIN Reality will be able to iterate with it.

One new possibility that several technology companies are looking at is Augmented Reality. Rather than replacing the real world for the virtual, AR adds an information layer to the game. One company I spoke with said that it anticipates having glasses that would pick up spin out of the pitcher's hand and transmit that information to the hitter in real time. That would require an incredibly quick transmission and reaction, but in theory, it's possible. The hard part is getting it down to a usable format and even then, a system like that is likely to be banned.

A more likely application for such AR glasses would be for pitch recognition training. There are systems now where small balls are fired out with colors and numbers so small that most people squint to see them, but when an elite hitter stands in the batter's box, they can call out the numbers as they go by. This is useful in situations wherein a hitter can't hit. Imagine a player that had shoulder surgery and is rehabbing, but doesn't want to lose his eye while he's trying to get his strength back.

In 1993, I attended spring training for the Minnesota Twins down in Fort Myers. One of the drills I saw involved several players standing about 20 feet back from the plate and firing squash balls—small black rubber balls—as hard as they could to a hitter. It surprised me how easily the hitter could adjust from hitting a baseball at 60 feet to a tiny squash ball from much closer, but they did.

It was one lesson I learned but about a decade later, I was at spring training with another team who had teed off on a rookie Tim Lincecum, then one of the top prospects in the game. The day previous, the team had knocked Lincecum around for four runs on a couple doubles and a homer. I asked one of the players if he thought Lincecum was overrated and he shook his head. "No, kid's real," he said. I asked why they knocked him around and the player said "He was tipping." We ended up in the video room where he showed me how Lincecum was lifting his index finger slightly on his curveball, letting the hitter lay off and force the fastball. It took me a couple slow motions to see it and then in fast motion, I blurted out, "You can't see that in time to do anything with it." As the player—a longtime infielder who no one would describe as a great hitter, but nonetheless a major leaguer—just looked at me, blankly. I realized in that moment that even the ones that weren't superstars were freaks. Yes, he could see that tiny motion of a pitcher's finger and yes, he had.

Something like WIN Reality allows baseball players, who already have superhuman vision, to train that vision on things that are very close to reality, over and over, controlling the situation physically but giving the proper mental and kinesthetic cues to make the training worthwhile.

As virtual reality continues to evolve, we're going to see better and better systems that feel more and more real. WIN Reality exists on a consumer platform, available on Amazon for a couple hundred bucks. Add in a twenty-dollar-a-month subscription and anyone can have a system that was almost science fiction a decade ago. Anyone, anywhere, is able to have one of the most advanced hitting systems possible. Given another decade plus the inevitable advances, it will be interesting to see where science and technology takes hitters. The next Mike Trout is likely to have been built virtually.

We started this chapter with a discussion of home runs. Home-run distance is one of those things that often falls to mythology. Mickey Mantle hit it here and Babe Ruth hit it there. Every minor-league stadium has a tale, but few have science, though that is rapidly changing. Perhaps that might take a bit of romance out of the game, but for every poet who doesn't enjoy it, there's a scientist who does. There's no reason that mythology and science can't co-exist. We'll never know where that Mickey Mantle home run landed, or if Ted Kluszewski really hit it that far. We can't even know exactly how far Glenallen Hill's famous home run into a window across the street from Wrigley Field went, even with video evidence and existing buildings. We can come really close, but not exact.

However, there's a point, largely defined by Greg Rybarczyk's work, where we can pinpoint the distance that Pete Alonso homer went or how far Shohei Ohtani sent it. We're learning exactly why the ball went that far as well, a combination of bat speed, bat structure, wind, air pressure, and who knows what else. No matter these details, there always seems to be a bit of magic sprinkled in there as well.

We can't know how far Vladimir Guerrero Senior's home runs went, but we know exactly, to the inch, how far it flew for Vladimir Guerrero Junior. That's progress.

4

PITCHING

I am often asked if pitching is unnatural. This question always strikes me as poorly constructed, but it's the way most people ask it, so here we are. To me, this question is much more about understanding what pitching actually is, what goes into it, and how to control for all the various forces, actions, levers, and fatigue that makes it so complicated and prone to failure.

Let's make no mistake of it here—modern pitching is a failure, resulting in nearly a third of all major-league pitchers sharing the same triangular scar on their elbow, the result of not controlling all those forces, actions, levers, and fatigue.

Let's break this down into component parts first.

Throwing—the act of picking up an object and propelling it with the arm—is far from unnatural. Anyone who's been around a toddler knows they can often have a pretty good fastball, or just toss that plate of spaghetti a surprising distance. No one taught them this. It progresses from grasping to lifting to propulsion in pretty quick order.

Early in human evolution, there's evidence that strong, accurate throwing was a competitive advantage. Throwing a rock at a rabbit is only effective if it's accurate and hard enough to at least stun the target, if not kill it. There's also question about whether long-distance hunters, who would basically run long distances to wear down their prey, began giving themselves more of an advantage by throwing rocks and/or pointy sticks. The motion for pitching a baseball and a javelin is very similar, mirroring the rock and the spear of our ancestors.

There's also evidence that the part of the brain that does the quick calculations and trigonometry necessary for accurate throwing—how much do I lead that rabbit if he's hopping fast?—contributed to the development of speech. The Throwing Madonna theory, popularized by neurologist William Calvin, posits that the area of the brain that does both was developed earlier by the increasing use of throwing in hunting.

Development of throwing in war continued on, with Greek hoplites using an overhand motion with their spears and Roman cavalry utilizing a similar technique on horseback, applying the force of their pony's run to impart extra force to their throws. Discussion of how new recruits were integrated into Roman legions by Polybius focuses on the need for training with all weapons, including spears. Most interestingly, this discussion seems to recognize the need for physical recovery and decreased effectiveness after heavy usage.

So the answer to the first part of this—is throwing unnatural?—is clear. No, it's a completely natural thing, rewarded as a necessary skill for centuries.

By the time baseball as the modern sport we know came around, throwing rocks and spears had been traded for machine guns and artillery. People still threw things, but they didn't do so as a matter of survival or even recreation. The early games showed this, with no distinct technique of overhand or underhand and no deception, basically just a way to put the ball in play. Pitchers were something just shy of a tee and their competitive instincts soon changed the game.

In his seminal work, *The Physics of Baseball*, Robert Adair explained why a curveball curves and for the next few decades, people debated whether Adair was right (he was) or whether a curveball was an optical illusion. It's hard to believe in today's data-driven age, but Adair was in a time of expensive calculators, not cheap computers and cameras.

Dizzy Dean, the 1930s pitching ace, knew better. "Go stand behind a tree and I'll hit you with an optical illusion," he told a journalist.

The fact is that curveballs, sliders, and really all pitches are subject to the laws of physics, especially aerodynamics. Within certain parameters, there's only so much a pitcher can do to move a ball, but it can move a lot. Just the act of measuring it seems to have changed the game.

Rapsodo, the camera technology that can capture and quantify the spin and shape of a pitch, changed the game like nothing else before it and the implications are still being learned.

Adair used a 1959 paper by Lyman Briggs in his book, showing that the expected arc of a curveball was between 10 and 17 inches, just as we find it to be today. The range is a bit wider, though as you learned in Chapter 1, that may be largely due to ball variations.

The simple explanation is that spinning a baseball creates aerodynamic forces and there's only so many ways that a human can throw the ball and make it spin. This is sometimes referred to as the "Pitching Clock." This isn't the countdown timer used at some levels of baseball, forcing a pitcher to throw the ball more quickly, but a theoretical construct of two elements of pitching, arm angle and spin direction/intent.

What we lack is the standardized vocabulary that allows us to discuss all of these with any measure of precision. Just look at a curveball—there's an endless number of nicknames for it, like Uncle Charlie, Yakker, the hook, and who knows how many more. Those are fine, but ask someone what a curveball is supposed to do and you'll get as many answers. Is the person you ask describing a "hammer curve" that moves up and down ("12 to 6") or a looping, oft-hanging curve that spins and doesn't have the late sharp movement that fools hitters? Is it a slurve-like, sideways motion, or a late break, where the ball simply jumps? All have been described as curves, but pitches can easily be described by the pitching clock, so here's my attempt to standardize it. If you can understand these, you'll likely understand more when people speak about them in the general terms they always will.

Note: In all of these descriptions, I will give the perspective of a right-handed pitcher. For a lefty, everything is opposite, including the clock and the movement.

CURVEBALL: We will start at the 12 o'clock hour on the clock. An ideal curveball has a spin that goes over the top and forward (toward the batter, away from the pitcher). The ball is spun hard over the fingers and aimed to the top of the strike zone. The ideal curveball will have a late break that goes nearly straight down ("12 to 6").

SLURVE: Here between a curve and a slider is a simple variant. It may be a bad version of the classic intent curve or slider, or it may be an intentional variant to give a batter a good look. More often, it's a description of result not intention, and moves slightly off the norm ("1 to 7").

SLIDER: Some will say that a slider is just a poorly thrown curveball and while that may be where it comes from, since pitchers often can't develop both and the natural movement will often guide pitching coaches on which flavor to work on, it has become a distinct pitch. While there's some variation on the movement from pitcher to pitcher due to spin and arm angle, the classic modern slider moves down and away from a right-handed pitcher to a right-handed hitter ("2 to 8"). There's also some variation in depth to the pitch. An early break brings it across the front of the plate, while a late break brings it to the back. You'll hear these called "front porch" and "back door" for obvious reasons. There's also a late and low break version that you'll hear as a "back foot" slider from a righty to lefty, or lefty to righty matchup.

IDEAL GYRO: This pitch exists in both theory and reality, but is seldom thrown because of anatomy. Imagine a ball with a perfect side spin, thrown like a football spiral. It is exceptionally difficult to get the hand on the side of a baseball and throw it hard while also pulling down. It can be done with a football quarterback's motion, clearly, but with a baseball, from a mound, very hard. The ball will move sharply to the side ("3 to 9"), but too often there's not enough spin on the ball to move enough air, which makes the pitch just steam forward like a flat changeup.

SINKER: Baseball is a crazy game, and if you ask most people what a sinker ball does, they'll describe a ball that drops straight down. No, that's a curveball. Despite the name, a sinker is thrown with pronation. (To visualize pronation, imagine holding a soda can in your hand. Now imagine pouring it into a glass. That thumbs down turn of the hand, with the top of the hand going to the glove side, is pronation. You should notice in that same motion how much both the elbow and shoulder are forced to alter position, which is why there's usually more stress on the arm from this pitch.) The pitch darts down and in to a right-handed batter. A hard sinker will induce ground balls more often than not, but not because the ball drops, as many would expect. It's because the ball moves in and down, away from the "sweet spot" of the bat.

SCREWBALL: This is a bizarro version of a slider and a more pronounced pronation with the hand. In an ideal world, this would simply be a grip-reversed curveball, with the ball coming out and dropping down. Instead, anatomy works against the pitch and leaves it as a harder-moving reverse slider, moving more laterally than pure drop ("10 to 4"). As with the sinker, this requires the shoulder to move more and can cause some issues in the elbow as well. It is almost exclusively thrown by lefties in the modern game, which essentially acts as a right-handed slider to left-handed batters. A pitcher that could control both, moving the ball in and out to both sides, would be fearsome.

There are no pitches that move "up" the clock, absent an unusual delivery like a submariner or true sidearm delivery. In fast-pitch softball, pitches that move up are common, creating even more deception and indecision for hitters. It's no wonder that pitchers truly control the game in high-level softball and no-hitters are relatively common. They use the full clock!

All of these pitches break, or move, and can all rightly be called breaking balls. Knowing the difference between each type of breaking ball is key to discussing pitching in a scientific context, so you're now way ahead of the game. Let's move on to the pitches that are fastball-based. Remember, that doesn't mean they don't move!

FASTBALL: Simple, right? Throw the ball real, real hard. Except that's not really how a fastball works. There are two main variants, the two-seam and four-seam, named for how the ball is presented to the wind. (I love that phrasing.) A fastball is thrown straight, with an ideal spin that is backward (as observed by the pitcher). If held so that four seams are seen spinning, it will get a little more aerodynamic force acting on the ball and it tends to be very straight in both planes, though a hard four-seamer will often appear to rise. Studies have shown and pitch trackers have confirmed that this is largely an optical illusion, dropping less than expected based on the plane of the throw. A two-seam fastball gets less aerodynamic force and can run to the arm side, while also being slightly slower than an equivalent four-seamer.

CUT FASTBALL/CUTTER: A cut fastball is not a ball that has been cut. It's confusing, but describes the motion, and for some reason "riding fastball" didn't catch on. This is basically a fastball where the fingers offset slightly from directly behind the ball, inducing a small amount of off-axis spin. For some, this is a flaw and for others, like Mariano Rivera, it becomes the Hammer of the Gods. Thrown well, there's late life and can be significant movement. There's less spin so it looks like a fastball to the hitter until almost to the plate, when it darts away (from a right-handed batter).

SPLIT-FINGER FASTBALL: The split-finger, or splitter, is something of an odd-duck pitch. It relies almost entirely on the grip, which is really just a placement of the ball between the index and middle fingers and is thrown like a fastball. Because the fingers are to the side, there is less velocity from a normal arm action and the ball has significantly less spin. It tumbles at the end, often dropping sharply. There was some thought that splitter grips put more pressure on the elbow, but there's no scientific evidence for that. The splitter is an evolution of the forkball, with similar action. The variation and plane of the split is dependent on the release of the ball. The closer to vertical the fingers are held, the more vertical (opposite) the drop will be.

FORKBALL: The forkball came before the splitter, but it's been largely abandoned due to the relative ease and similar action of the splitter. Roger Craig, the longtime Dodgers pitcher and Giants pitching coach, is largely credited for the variance. He was known to use the forkball, but taught the splitter. A forkball also drops, but the grip is deeper, jamming the ball in between the fingers. Pitchers will often practice with softballs or even shot puts to train their fingers to handle the ball. The forkball is also thrown with a sharp wrist snap. Again, this was thought to be harder on the elbow, but mostly, it's just harder to synchronize the wrist action and get the pitch to work.

CHANGEUP: Here's where things get complicated on what should be a simple pitch. Instead, we have almost endless variations on a simple theme. A changeup or "change of pace" is a pitch that is thrown with the same arm action and, notably, the same arm speed as a fastball, but the grip creates resistance and reduced spin, bringing the ball out at significantly reduced velocities. Most will try to keep their changeup about 10 mph slower than their fastball for the best effectiveness. There are likely infinite grips for a changeup, with the most popular being a circle change, a palm ball, or a "vulcan" fork, where the fingers form a V between the middle and ring fingers. Most modern changeups are thrown with a bit of pronation, especially the circle, which is thrown with enough pronation to be considered a variant on a sinker.

There are other pitches, but they're mostly variants of these types, with only the no-spin knuckleball as a significant variant.

Intent and Mechanics

Once we get past the pitch types, there's another issue—intent versus result. A pitcher on the mound gets the sign from his catcher, who puts his glove where he wants the ball. The intent is clear, but the result is often different. The ball doesn't break enough, or breaks too much. The pitcher throws to one location, but hits another. Do we classify pitches by intent or result? For systems like Hawk-Eye and Trackman, they only see the result. Intent is what the pitcher does or tries to do, but we don't see that prior to the pitch (unless you're cheating).

There's also a combination of factors, such as arm angle, that affect things. By dropping down the arm to a three-quarters angle, a "12 to 6" spin essentially changes to "2 to 8" without any other changes. The pitch result would look more like a slider than a curveball despite the intent, while a sinker would have more downward movement. This varies from pitcher to pitcher because none are the same, but within a single pitcher, at least at high levels, they tend to release the ball from a similar point, which makes it harder for a batter to pick out what pitch is coming.

This concept, known popularly as pitch tunneling, was first discussed as a very specific perceptual point, about 23 feet out of the pitcher's hand. Combined with what we know about hitter reactions and decision speed, having a pitch come from a similar location makes sense. Even the simple changeup is based on the arm looking fastball and the ball coming out slower, ruining a hitter's timing.

Tunneling remains controversial, partially because it's largely strayed from the original intent. Instead, it was overtaken by the concept of pitch design. A series of events, culminating with the Rapsodo device that could track ball movement and spin at a relatively reachable price point, allowed pitching coaches to offer a new kind of coaching. Rather than simple trial and error, they could work with pitchers to refine their arsenal, shaping the pitches to move this way and that.

That leaves the rest of it to the pitcher. Arsenal designed, situation factored, and the pitch still has to be executed and even then, a good hitter will hit a good pitch some percentage of the time. Measures of pitch quality are lacking, again because outside the teams themselves, it's hard to know the intent. Even inside, very little has been done with this. A team analyst that works largely on knowing and combating pitch design says that pitch quality is simply not worth knowing. "It's interesting to a few, I guess, but the result is all that really counts. They don't pay me to find moral victories, just Ws."

That means that pitch type, pitch execution, and pitch perception are all things that are controllable by the pitcher, in addition to deception and timing. To go beyond that, we have to look beyond the pitch itself and to how that pitch is propelled.

Starting in the mid-2000s, MLB was able to show data about the pitch beyond simple speed. They could show pitch shape, spin, and a number of other factors, including pitch type, which was a simple result-based system. If the pitch looked like this, we call it that. With a Rapsodo unit, first introduced in 2017, which uses a special camera to track a pitch, showing its spin rate, direction, and the flight of the ball through the full distance from the pitcher's hand to the plate, that information became available to anyone at a cost lower than a small car.

What remained beyond normal usage is that last real piece of the equation: biomechanics. For years, pitchers had been able to go places like the American Sports Medicine Institute in Birmingham, Alabama, pay a couple hundred bucks, and suit up in what one Mets pitcher described as "underwear, way too tight" with markers all over it. Using a million-dollar system, huge computing power for the time (you know, a Compaq 386 with a four-megabyte hard drive), and a lot of brainpower, Dr. Glenn Fleisig brought the science of biomechanics to the art of pitching.

Attempts had been made previously, but there was a confluence happening in Birmingham. There was an opportunity for Dr. Flesig to work in partnership with Dr. James Andrews, one of the top sports medicine surgeons in the world. The pitchers that came to Andrews for surgery could then be checked for biomechanical issues, with the hope that someday, pitchers would come before they needed surgery.

"ASMI was my attempt to put myself out of business," Dr. Andrews told me in a 2006 interview. While that didn't happen, it did help slow the acceleration of injuries that was being seen. Though injuries have trended up over the last twenty years, the slope is not as steep as expected. Rather than preventing injuries, the improvements in biomechanics and all systems of sports medicine and sports science have only been able to hang on to the reins and slow the losses.

Part of this is the acceleration of injuries at the lower levels, where the technologies and systems of baseball, as well as the quality of coaching, are much lower. Unfortunately, even for the top-end destination ballfields that are outfitted with camera systems like Synergy and Trackman radar ball tracking, as was used in MLB until 2020, there's not a single one below the major-league level outfitted with Hawk-Eye, KinaTrax, or another system that is similar.

While there are more places around the country that have biomechanics labs, including Driveline's facility outside Seattle, these are niche facilities and aren't widely used. They're not just a 1 percent technology, they're a .1 percent technology. There simply hasn't been time for some of the technologies to trickle down. They've only been in the majors a handful of years and even then, major-league teams are having such a hard time with that data that they're going to outside consultants to help.

That's where one Canadian company comes in. Dr. Mike Sonne is the Chief Scientist at ProPlayAI and what they've done is take all the data of multi-camera systems and biomechanics labs and made it functional as a phone app. Instead of a multi-million-dollar fixed system that requires tons of setup, one can simply take their phone out of their pocket, anywhere, anytime. It's a promise that's almost too good to be true when you first hear it, but the technology is both validated and usable.

I sat down virtually (Canada was still on COVID lockdown . . . hopefully those of you reading this in the future have to look that up) with Dr. Sonne where he showed me the ins and outs of the system, from arm speed to hip/shoulder separation. He explained how he created a system that simply wasn't possible years ago. If you think back to Dr. Fleisig's first lab in Birmingham, he had about 1/1000th the computing power of your average iPhone or Android.

Dr. Sonne explained to me that what biomechanics does is essentially function as a radar gun for the body. "Much like we see one pitcher come in and throw 90 mph but still be successful in a league where velocity has become king, players with untraditional mechanics can still be successful," he said. "The science of biomechanics is about measuring movement, just like how ball tracking measures velocity and spin rate."

By knowing this kind of information, there's a lot to be gained, but that hasn't been available to many before a tool like ProPlayAI. "Much like with ball tracking, a pitcher needs to know about all of the characteristics of their movement to maximize their potential," he told me. "These numbers can give insight into if a pitcher is not utilizing the ground effectively, or transferring the energy from their legs to their arm. These metrics, including arm speed, can also give a pitcher insight into if they are changing their mechanics to throw different pitches—something that could be a tell for skilled hitters."

In other words, there's a good and a bad, a yin and a yang to biomechanics, but it's also objective. A pitcher might have numbers that indicate there's a problem or that there's potential. There's also the ability to measure arm speed that functions on both sides of the game. A pitcher needs to have his changeup look like his fastball, as explained in this chapter, but previously the only way to measure this was either to be a major-league pitcher or to find out, as Earl Weaver would often say, "when the hitters let you know."

Sonne understands that a lot of major-league teams have similar data, but there are limitations. "Markerless motion capture in stadiums and capturing every single pitch is going to give us a world of insight that was never before possible. Most pitching biomechanics research has been conducted in labs, in environments where performance was not prioritized over data quality. That's not so much a criticism of the labs as much as it is a reality," he explained. "The teams that have been on the leading edge of using these systems are very confident in what they are delivering from understanding what pitchers have optimized their mechanics. They also know a good deal about what their pitching staff is capable of changing—meaning if they scout someone that does a lot of things well, but perhaps doesn't have a firm enough front knee, they know they are good at changing that internally and could look to acquire that player in a trade."

With this kind of data available, can teams also get an immediate boost in terms of performance and injury reduction? Sonne not only thinks so, he has the data to show it. "Performance and health are so tightly linked, that it can often be impossible to split the things that lead to high performance from the things that are attributed to prolonged health. The only truly agreed-upon data that has a clear link to injury is pitching while fatigued. Fatigue shows up in a lot of different ways, but velocity drops usually only occur after the pitcher has changed their mechanics to maintain their performance, while sacrificing how they move. Our research has indicated that mechanical changes occur before velocity decreases, and we've published it."

Sonne's hope is that democratization of data comes from an approachable cost. More data helps create better models, which creates more insight. It's a cycle that really does bear out the promise of artificial intelligence in baseball, paired with an accessible system that can help as much in a major-league bullpen as a sandlot in Brooklyn or a cornfield in the Dakotas. It's that increase where Sonne's system really comes alive.

"As great as the in-stadium markerless capture units are, we know that close to 50 percent of the total throws a player makes are not in games," he said. "We also know that biomechanics data have traditionally been very expensive to capture. Finally, we have known a lot of the risk factors for injury from a workload and velocity perspective, but the mechanics part has been hard to capture—due to the expense, or to the inaccuracy, of subjective

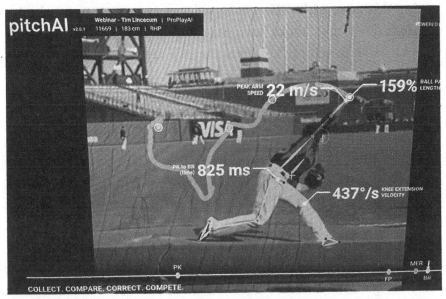

Tim Lincecum, biomechanics. Data from ProPlayAI/PitchAI.

'coach's eye' methods. With that in mind, we have wanted to explore how advances in computer vision technology could be paired with sound scientific methods to reduce the barriers for data collection in biomechanics. We are very happy with how far we have come. With close to 50,000 pitches analyzed, we're getting very close to delivering insights into what mechanics really mean for player development, health, and scouting."

Getting from 50,000 to 50 million is a long journey, but the availability of this kind of technology reminds us just how far we've come and how far we have to go. I asked Sonne what he thinks is coming next in the field. "With so many barriers being removed around requirements to capture good quality biomechanics data, we are on the verge of a big data era in human movement science," he said. "Apple's AR kit and many other accessible technologies make the ability to capture data at scale so much more accessible. Much like we are seeing detailed statistical leaderboards on pitch characteristics included in every scouting report, biomechanics data will feature prominently in scouting reports. We're already seeing this now, with different broadcasts using our technologies to break down mechanical advantages that some players have over others. I like to tell my students that we are in the golden age of kinesiology—it has never been easier to collect human movement data. Now it is time to figure out what to do with it all, and I feel we're on the verge of exponential understanding."

Sonne isn't just a scientist, but also a fan, who regularly says he grew up in the SkyDome (now Rogers Centre). I asked him what he notices when he first sees a pitcher and is it as a fan, or as a biomechanist? "The first thing that hits me every time I see an elite pitcher throw, is just how impressive the human body is," he responded. "The internal rotation of the shoulder is one of the fastest movements in all of human motion, across any sport. As a kinesiologist, biomechanist, and fan of baseball, there isn't a time that I watch a pitcher and am not blown away by how difficult, humbling, and beautiful the sport of baseball is."

A technical bent and a love of the game is a nice sweet spot for anyone trying to make the game better. What Sonne and his team have done is nothing short of astounding, bordering on the Arthur C. Clarke statement about a sufficiently advanced technology being indistinguishable from magic. What Sonne has here is what baseball needs—a little magic for pitchers.

(Oh, he's working on the same technology for hitters. I guess they need some magic, too.)

One of the questions I'm often asked is how pitchers from previous eras put up seemingly huge workloads in terms of both innings and pitch counts, but today's pitchers break down at far lower numbers. The fact is, there are two major factors.

First, we largely remember the outliers—Nolan Ryan and Tom Seaver are Hall of Famers for a reason—but at the same time, look at the hitters of the era. Ozzie Smith hit his "go crazy folks!" home run that's famous, but few remember that was his only homer of the year. Smith, a no-doubt, inner circle Hall of Famer, had 28 career homers. Dave Concepción, the shortstop at the heart of the Big Red Machine, never hit more than 16 in a year. There are two shortstops in 2021 that had more than that ahead of the All-Star break! There's no batters to "ease up" on anymore. In 1905, Christy Mathewson wrote a book called *Pitching in a Pinch*, where he espoused letting bad hitters put the ball in play, but saving one's best stuff for the best hitters and situations where runs could score. Even Ryan did that, regularly changing the speed of his fastball from the high 80s to mid 90s, but occasionally throwing that 100 mph heat that the hitter always had in mind.

Second, we're likely looking at the wrong numbers. Innings and pitches tell us next to nothing. At lower levels, like Little League, a pitch count is often the most accurate measure possible, and we've seen an increase in injuries for travel teams and the like that don't have any limits. One better measure is called acute to chronic ratio and was developed by Dr. Ben Hansen for pitching-specific monitoring. Originally developed in Australia by Dr. Tim Gabbett for sports like rugby, Acute to Chronic Ratio (ACR) is exceptionally useful and accurate in studies. We can even use it to look backward.

The Case of Nolan Ryan

Thirteen innings. Nineteen strikeouts. Ten walks. Nolan Ryan's outing on June 14, 1974, was a stunner, especially given the pitch counts of modern baseball, but even in 1974, it was an extreme outlier. Pitchers simply did not go 200 pitches, even in the "single pitcher" era when complete games were the rule rather than the exception.

Ryan is, of course, an outlier in almost every way, but what lessons can we learn not only from this exceptional outing, but his 1974 season? Variously reported as 235 and 238 pitches, almost everyone can agree it was a lot.

According to several studies, the ACR is key to understanding and managing workload. Dr. Ben Hansen helped me dig into the numbers from 1974 to see if modern workload management can be used to explain Nolan Ryan and give us lessons on how to manage pitchers now.

First, there was some math. While 1974 isn't in the distant past, we don't have accurate pitch counts for players in this era. That forces us to use the pitch-count estimator rather than accurate pitch counts. It's shown to be accurate enough that we feel using it is acceptable for a format like this. (Ryan's noted 238-pitch game was estimated at 242.)

Hansen then ran those estimated counts through some calculations and created this chart, which shows the season-long ACR for Ryan:

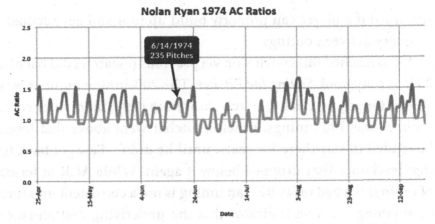

Chart of Ryan's A:C Ratio created by Ben Hansen/Motus Global.

Surprised? I was as well. Ryan is seldom outside the "safe zone" between 0.8 and 1.2. He's regularly at the high end of this range. The spikes outside the range are often followed with short outings and those short outings often seem to work against Ryan's health, which is counterintuitive.

In fact, Hansen highlighted one part to me, showing how the rest period for the All-Star Game worked against keeping Ryan's ACR ratio in the safe zone. There are also a couple of low-count outings of 73 and 28 pitches that had noted effects on his ACR. (That 28-pitch outing came on two days rest after a 158-pitch outing!)

Yes, that does indicate that if a pitcher has a quick outing, they should either "complete the workload" in the pen or be made available in relief during the following days. This is again counterintuitive, especially given modern protectionist methods, but the underlying science of workload management is well proven in many areas. It remains to be seen how well it will work in pitching, but early studies indicate a major reduction in arm injuries, even in counterintuitive scenarios.

There are likely some upper limit to this. Even Nolan Ryan couldn't go 200 pitches time and again, but a high workload can be sustained if a player can properly build up to it and get adequate recovery between outings.

I also reached this conclusion via previous research I did back in 2005 for my book *Saving the Pitcher*. The 200-inning number was a need for any ace, but research showed that the more important number was 190 innings. When a pitcher went above that level, he tended to stay there for years, until he didn't. Few got back to that level once they dropped below it again. While ACR in terms of innings is a bad measure—an inning is not a consistent measure of anything—it is an indicator that the underlying real measure was being suggested in other ways, like inning limits.

There are a few caveats to this that, even as a pure mental exercise, should be pointed out. First, the pitch counts are estimates. While reasonably accurate, having actual throw counts would be significantly better. Second, these are *pitch* counts only. Every throw counts, so we don't know how Ryan's side work, which became noteworthy later in his career, would have affected these ratios. Third, the early-season ACR being outside the safe zone should be ignored. Ryan had a spring training where he was throwing and building up his arm, which can't be included in this example.

Of course, we shouldn't discount how unusual Nolan Ryan was. (Note: I reached out to Ryan for comment, but he declined.) Ryan did things few can do, but perhaps the durability was more than just being a genetic outlier. Few remember that Ryan did eventually break and had elbow issues throughout his career.

These noted, it's clear that Ryan's 1974 season indicates that heavy workloads—much heavier than what we see in 2018—can be handled safely. Not everyone can be Nolan Ryan, but many pitchers and coaches could learn an interesting lesson from Ryan's workload and workload management.

Conclusion

Pitching isn't unnatural, and it's being pushed forward by technology. Nolan Ryan was a clear outlier, but we're seeing staffs full of guys with his stuff, with better training techniques, better sports medicine, and a willingness to use progressive methods. We're seeing pitchers today that are unquestionably the best of all time, throwing what Crash Davis described in *Bull Durham* as "exploding fastballs and ungodly sliders." Pine all you want for Bob Feller and Nolan Ryan. This is the golden era of pitching, and one of the reasons is the support and push given to all of them by science.

5

THE FIELD

Football fields are boring, whether the American or European version. They're basically big, flat rectangles with markings. Basketball courts are all the same in dimension, despite the best efforts of some overzealous marketing staffs. There are some dimensional variants in hockey, but it's a couple feet here or there.

Baseball is different.

The Green Monster. The short porch in right field at Yankee Stadium, new and old. There was a natural hill they didn't flatten at Crosley and a stupid, artificial hill in Houston. There's nooks, crannies, materials, padding, baskets, and any number of other variations, and those are just the fences, let alone the distance to them.

For Major League Baseball, there are rules about just how close the fences can be, but everything is pretty much a suggestion in terms of dimensions. There's nothing stopping the next new park from being like Griffith Park or the Polo Grounds rather than the latest 335 down the lines, 400 to center that became standard in the Selig Ballpark Era, as architects worked more on how close they could get the fans to the experience without thinking much about the experience on the field.

The trend is more noticeable in the minor leagues, where anything unusual has been pushed to a more updated but more neutral style. The upgrade for many towns around the minor leagues was notable (and expensive), but few have architectural details that would be rated as charming or even notable. Instead, they more and more fought to be what is unromantically called "park neutral."

That term largely comes from Clay Davenport. Davenport, a meteorologist by trade, was one of the founders of Baseball

Prospectus, the sabermetric powerhouse of the 1990s and early 2000s. Davenport's work was largely based on the idea that comparing players had to be apples to apples. Players played in different parks, different leagues, and in different eras, but he worked to be able to translate all of those into a neutral context.

Instead, Davenport transformed the game. While Bud Selig is given credit for pushing more and more teams to build new stadiums on their tax dollars, Davenport's work basically created the template for how those parks were built. In essence, they all sought the Davenport ideal—a perfectly neutral stadium, often at the expense of character and home-field advantage.

Prior to Davenport's work, parks were often fit to the contours of the area, the way that older parks like Fenway in Boston or Wrigley in Chicago are shoehorned into the neighborhood. The Green Monster is only there because even back a century ago, prior to the Babe Ruth home-run onslaught, they knew that a short distance could be augmented by a higher wall. The modern equivalent is PNC Park's right field, which is seemingly there because of the river, but the ballpark is set back so far as to not be a "real" issue. It makes for a nice story, at least, and a beautiful ballpark setting.

Davenport's work often described parks as "pitcher's parks" or "hitter's parks." This became almost a pejorative, especially in the case of the California League, driven almost to extinction by the extreme hitter's environments. One park, in Lancaster, California, was such an extreme hitter's park that teams were reluctant to send players there and numbers in the league were skewed. A pitcher with any measure of success was hailed, but so few were successful that eventually, teams pushed to add teams to other leagues just to avoid Lancaster and the like.

Today's major-league parks all play to the neutral, with only a plus or minus nine from the neutral (usually expressed as 100). Surprisingly to most, Coors Field is a neutral park now, largely due to the use of a humidor to keep balls moist in the high, dry environment at a mile high. On the extremes, if you can call it that, are Atlanta at 109 (9 percent higher than neutral advantage for hitters) and Seattle at 91 (9 percent lower for hitters from neutral).

Of course, park factors aren't as simple. Some parks are extreme for home runs, others for triples, which usually indicates deep or strange dimensions. For homers, Yankee Stadium leads the league in 2021 with 128 in homers while the low is 68 in Detroit. Triples might be the most extreme of the extremes, where Houston has a 321 park factor for the most exciting play in baseball.

Park factors even have differences when it comes to walks and strikeouts, though these again tend to fall in ranges and seem to fall more to lineups than to some quirk of the hitting environment.

Still, through the so-called Selig building boom that traversed the so-called steroid era through a home run–heavy environment to the extremes of pitcher favor in 2020 and 2021, the parks have largely not been a part of it. Instead, they have gone neutral. That's an unintended consequence to be sure, but one where Clay Davenport deserves far more credit than ol' Bud.

The World Is a Baseball Field

Baseball fields were once just that. Fields. Teams would lay out the bases and then just… play. Manicured fields? Stands? No. Ol' Hoss Radbourn, the pitching great of the late nineteenth century, would often tell the tale of the time he ended a game with a homer. It didn't go over the fence—there was none—but instead went under a horse. One of the fans had just ridden up and when Radbourn's hit went near it, it almost kicked the outfielder in his head! (Would that have been an error, or an assist for the horse?)

Fields changed quickly, mostly for one simple reason: People would pay, and if people would pay, some wouldn't and had to be blocked from seeing. The people that would pay demanded comfort, a bit of safety, and a good view. The early stadiums had that, wedged into neighborhoods or built on the outskirts of town.

Those would change, grow, and modernize up to this day, but the dimensions have only changed to the fences. If there's one thing that's been the most consistent in baseball over all the years, it's that 90 feet between the bases.

At first, the baselines weren't marked or really measured. The bases were often stepped out or staked so that they could be measured once. The area between them was dirt, but ruts from where men ran, stood, or even slid. By the 1890s, almost all fields were enclosed on all sides, but the distances were far from standard, ranging from 150 feet to 500 in places, depending almost entirely on natural terrain.

Chicago's Lakefront Park was perhaps the shortest ever. Al Spalding spent $1,800 on the grounds, with a wooden grandstand that could seat 1,500, but the left-field fence was only 186 feet away. The biggest downside of this park wasn't its shape, but its construction. A wooden structure was a bad thing when the Great Chicago Fire hit. Spalding rebuilt, adding more grandstand space, bleachers, and even something he called "Sky Boxes," which had chairs and curtains. His box had a phone, so he could run his growing sporting goods business. The fence in the newly constructed park was still short, leading to a barrage of home runs by the home team. The team was forced to move in 1884 and the National League set a minimum fence distance of 210 feet.

On the other side of that was the Red Sox first stadium, the Huntington Avenue Grounds. Built where Northeastern University now is and across the tracks from the Boston Braves, Huntington was the home of the first World Series and of Cy Young. There's a statue of him near the field now, but that field was massive. Consider all of Young's numbers in the context of this—the field was 350 feet to left, 440 to left-center, and 530 to center. Right was a more reasonable 320, but there was a tool shed in right-center that was in play.

Those extremes are gone now, left to this kind of "Wow, what would that look like?" given the lack of pictures. The early movies we have are choppy affairs, mostly done with hand cranks and seldom at games so that the videographers could get close for both lens and lighting.

Today, there's not so much a groundskeeper as there is a turf scientist. The strain of grass is hybridized and picked, often grown off site and brought in as specialized sod in trays, paired with a prepared substrate that allows for quick drainage and even suction beneath the stadium. Rainouts be damned on a modern field that ol' Cy Young could only have dreamed of. The dirt is no longer brick dust, bought from a nearby mason and instead is a specific mix, wet and tamped, raked and tractored into near perfection. No rocks, no bad hops, no unintentional quirks.

Even the grass is no longer a guess. There are several teams that model out how the ball will move in both the infield and outfield grass, which allows them to tailor it in the same way golf courses do for players. Want it a bit slower? That's easy, just cut it this way and roll it to push the ball in. Want it faster? There's an easy solution for that, with the field able to be kept at a very consistent "track" if they want it that way.

Using these methods, it's possible to adjust for different lineups and pitchers, but most teams don't do this. The consistency is often more important, especially if in-game changes are made. As well, a team can cut grass all it wants, but making it grow the next day is tougher, even for grass as closely tended as what you see on a major-league team.

While artificial turf is almost gone in MLB, the teams that use it also alter with preparations. Most of that happens with the substrate, which is the layer of substance under a field. It's usually seen as a black rubber dust, which can kick up when a player or the ball moves across it. Several teams, including Tampa, have shifted to a coconut substrate, which puffs up white. Teams can make the surface bouncier or harder, though there's not as much range as can be created with grass fields.

In fact, that grass has become a science unto itself and has changed the game in ways that many haven't even noticed.

The Green, Green Grass of Home Field

Many of us mow our grass the same way our parents did, trudging back and forth while a loud, gas-powered blade spins just a meter from our often unprotected feet. Maybe you're fancy and have one of those zero-radius riding mowers, or just hire it out. We all know—or are—that guy whose hobby has become having the best lawn in the neighborhood.

I'll bet that guy doesn't have a masters in turf science from a top research university like Purdue. I'll bet he can't look at the dirt and know just how to get a specially selected and grown sod to lay down perfectly, with imperceptible seams, and how to make it not only beautiful to look at, but functional in a way that most—even those that play on it—don't recognize.

Joey Stevenson is the Director of Field Operations for the Indianapolis Indians. If you think that's an unnecessarily corporate sounding title, you don't understand the job and frankly, I didn't either. After all, Stevenson does have that masters and his work on a field is often called the best in the minor leagues, a field that just celebrated its twenty-fifth anniversary.

Stevenson and other head groundskeepers have a lot of specialized training and work long hours to keep their fields in top shape, but who do people think of when you say "groundskeeper"? Carl Spackler, the gopher-obsessed Bill Murray character from the movie *Caddyshack*!

"Our profession has been hurt quite a bit from the movies *Major League* and *Caddyshack*. It's the first reference when people think about our profession. However, the systems we manage now agronomically are the Formula 1 cars of turfgrass management," Stevenson explained to me, sitting on the wall just steps from green grass that looks almost perfect, all the way to fence in left. "These fields are built and engineered to withstand hundreds of events per year, withstand seven inches or more of rain per hour, sensors measure evapotranspiration rates, soil moisture sensors, light meters, etc., etc. Our infield soils are engineered to perform based on the climate of a particular region. Over the last twenty years, our industry has moved forward quite significantly."

I asked Stevenson what goes into a typical game day for him and his staff. "A typical game day is the same almost every day, but we must take into account humidity, dew points, sunlight, and weather. Essentially we are setting up for a professional event seventy-something times per year (plus special events) and consistency is king for our players. The week before a homestand starts is where we start our prep. The grass is fertilized, aerified, mowed daily, fungicides/pesticides are sprayed, and our 'dirt' areas are prepped. The week before we are prepping the field to withstand the seven-game homestands we play."

That's a lot, but he's not done yet. "When the homestand starts, we focus on 'dirt' surfaces, primarily the infield skin. We start working deep moisture into the skin two to three days before the first game. We will flood the infield daily," he explained further. "Once the big day arrives, we use various tools to really work that top half-inch of infield skin to play consistent. We want the ball to bounce from grass to dirt back to grass and keep the same hop. We manage that with water and a tool called nail dragging. It's really a drag (pulled behind a small tractor) with 20D nails."

Done? Not hardly. "Then we must watch the weather. You can almost guarantee anytime it rains, the field will be tarped. Sports turf managers want control of the moisture in the skin and we can't always trust Mother Nature. It's common for us to water the infield four to six times a day on game day. Once we get that moisture we like, we must try and keep it for the seven days, but also watching radar and taking into account the weather."

I realized that in all of that, Stevenson hadn't even mentioned mowing. "The field is mowed every day. It's essentially a large golf green. Underneath the turf is ten inches of sand, followed by four inches of pea gravel. We do this so the field will drain and resist compaction, but with sand you leach nutrients very quickly as well as water. You will consistently find us spoon-feeding nutrients every two weeks, along with products to help keep the root zone moist, so we aren't using large amounts of irrigation. It's a balance."

As Stevenson notes, he has to work on a schedule as well. "The team is out working on the field each day around 2, so we must get our work finished by noon each day. Mowing, nail dragging, dragging, adding infield conditioner, repairing mounds and plates, cleaning and dragging the warning track, cleaning dugouts, etc. Add in that we face the same issues the world is facing, with a labor shortage. Some days we will have two people and some days we have four." Just thinking about this is tiring, and it offers a new level of appreciation of the green grass after all these years.

There are always stories about how a groundskeeper can do little tricks to make a field advantageous to his own team, creating a real home field advantage. Utilizing wetter dirt at first to slow down a fast-running team or harder dirt at second to have that fast runner slide over the base. There's a lip or even a slope that can send a bunt fair or foul. Stevenson thinks that's overblown for modern fields. Asked how much he can alter play, he answers "A little. The biggest difference comes from when the teams travel from the north to the south. In the north we grow Kentucky bluegrass mowed at 7/8". In the south, they play on Bermuda grass mowed at 1/2". The ball is quite a bit faster on Bermuda grass."

The height of that grass is remarkably consistent, again, constantly touched up like a top golf course. "Grass height is the easiest. Mowing taller slows the ball down and mowing shorter speeds up the ball. We can also spray plant growth regulators, which help the grass grow slower and more consistent. That spray will help keep the grass more consistent," Stevenson explained. "We can roll the infield skin really tight to speed up the ball on the infield. We can add more water on the infield skin and slow the ball down. We can dry down the grass to promote faster ball roll or we can have the grass a little more wet to slow the ball down."

While it might not be the old groundskeeper's tricks we heard of as baseball mythology, do these thing make a difference? "I think most of this is minimal at best and plays into that saying baseball is 90 percent mental. If a player thinks the infield is wetter and slower, he or she feels better about it . . ." and he shrugs.

One thing that you see in the majors, and often in the minors as well with their need for attention and promotion, is aggressive striping or even designs put into the grass by rolling and even painting. I asked Stevenson whether this affects how the field plays. "Yes it does," he answered. "The turf will get grainy and pull the ball in different directions. We try and always have stripes go toward the players in the outfield. That helps eliminate that. It's almost important to change up the patterns often. You don't want to roll the grass in one direction for too long or it will lay over and cause severe grainy issues."

Stevenson's clear—he doesn't like flashy. "I am not a fan of designs, because ultimately the grass isn't about what the groundskeeper can do, it's about what the players can do. Silly designs or large letters may look good, but ultimately provide no benefit for playability. How often do you see golf greens with a large letter or design? You don't, because it would affect ball roll."

Of course, probably the most common question that Stevenson gets is about the ol' lawn, but if you want your lawn or your local ballfields to look like Victory Field, you'd better start by hiring Stevenson. This isn't something just anyone can do. "This profession is highly specific and trained. I always compare it to a graphic designer—everyone thinks they can do it, but the best are worth it for their experience and knowledge," he explained. "We use engineered soils for our skinned areas and years of research and plant genetics to get grasses to work specifically for professional sports."

He continued, "I often get the question, 'how can I get my high school field to look like Victory Field?' My first question is, how much money do you have? You can have Victory Field without trained staff, but you need proper construction and engineering, draining, specific USGA sand, irrigation, engineered soil, and various clays and soils that come from all over the USA. Our infield is from Pennsylvania, our clay is from Iowa, our warning track is from New Jersey, our grass was grown specifically on sand in New Jersey from years of research and testing. Our indoor mound clay comes from Arizona and we have a crew of people who work seven days a week to keep this field in shape. Although a high school may want Victory Field, they really *don't* want it. It takes time and money and education to keep this field in shape. You may want a rain tarp, but do you have the staff to tarp the field over the weekend for a game on Monday? And a staff to remove the tarp once the sun comes out because it will kill the grass? Fields in MLB and here in the minors are comparable to Disney World: we have everything we need to keep it up."

Maybe Victory Field isn't the happiest place on Earth, but for fans who see it and players that play on it, Stevenson's field is as close as you come. Stevenson's work over the years has kept Victory Field immaculate and better than some major-league fields when it comes to surface. Opposing teams actually will send their players to rehab when their minor-league teams are the visitor at Victory because they know the conditions are near perfect. Stevenson and his crew control everything but the weather, but now, some are adjusting for even that.

Checking the Wind

The newest data point for parks is wind. We've known for years that wind has a big effect on batted balls (and even pitched balls to a lower extent), but a company called Weather Applied Metrics is taking this steps further. Their vice president of engineering Ken Arneson is a longtime friend and he explained the system to me.

"The effect of altitude is, of course, more consistent than wind. The typical fly ball at Coors flies 15 to 25 feet farther than at sea level. The longest home runs get somewhere between 40 and 50 feet extra distance at Coors. Every single fly ball there is affected by the altitude," Arneson said.

"Wind comes and goes, so most of the time, the effect is smaller than the effect of altitude at Coors. But when it does really get going, the effect of wind can be greater than the effect of altitude. The biggest effect we've seen is a fly ball at Wrigley that was hit by Patrick Wisdom, which was pushed back 78 feet by the wind, and also 34 feet to the left. It would have been a home run, but didn't even end up reaching the warning track."

Of course, it's never as simple as one variable either. They interact. "There was one home run hit by Chris Taylor at Coors that was pushed backward 40 feet by the wind, but got 38 feet added from altitude, so the wind and altitude sort of negated each other."

For years, Wrigley Field was known as a ballpark that was almost bipolar. Often, the first shot of a television broadcast or the first words out of Harry Caray's mouth were about the wind. If it was blowing in, the pitchers were going to be happy. If it was blowing out, then get the popcorn ready—the home runs would be flying out to Waveland Avenue.

In fact, if you look at the data Weather Applied Metrics has been able to collect, wind is far more complicated than in or out. Wind is a three-dimensional issue, blowing in and out, up and down, and everything in between, often at the same time. Everything from the architecture of the building, surrounding structures, and much more can affect how the wind behaves.

Weather Applied Metrics has gone as far as creating computational fluid dynamics models of the stadium, similar to what Formula 1 teams do with race cars. There's some stunning findings. Arneson showed me one model, where there's a flow of wind at the top rim of Oracle Park in San Francisco, home of the Giants, and that just below that, it's swirling in multiple directions. At the field level, it was blowing in to the hitter. For years, that would have been the gold standard, maybe tossing a few blades of grass in the

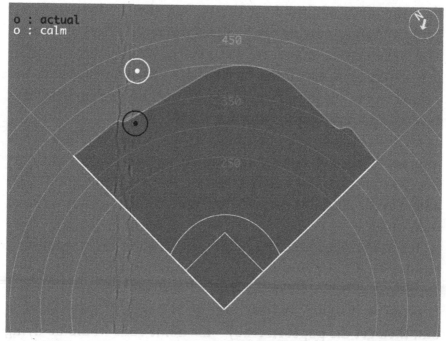

Patrick Wisdom's wind-aided homer. (Courtesy Weather Applied Metrics)

air, looking at the flags, and guessing. Now, Weather Applied and teams all over can do this on a pitch-by-pitch basis and learn even more about their own players and how to gain a real home-field advantage.

In the near future, teams may undergo design work, down to the overall design of the park, to maximize or at least attempt to control for these wind effects. "I know when the A's designed their proposed Howard Terminal ballpark," said Arneson, referring to the proposed new ballpark in downtown Oakland, "they ran some wind tunnel experiments using actual models of their ballpark for the Environmental Impact Report. But the purpose of the EIR is to measure the effect of wind on human comfort, not on fly balls. If some future ballpark architect wants us to calculate the effect of their design on the actual play of the game, we'd be happy to help them."

The baseball park is more than just where we go or what we see. It's not just a field, picked at random, and constructed. It's far more than "If you build it, we will come." Scientific principles go into the layout, the maintenance, and the playability of ballparks, from the grass of a ground ball to the air carrying a long fly. There's scientists building and maintaining the dirt, and scientists checking the airflow in real time, all so we can sit back with a cold beer and enjoy a game.

6

FIELDING

Is it a hit or an error?

This is the basic question of fielding, but if you listen to the modern masters, this question doesn't matter at all. It simply doesn't go deep enough and frankly, an error is a decision made by some person up in the booth with all the human biases and foibles, who may or may not be right. (They're mostly right in the age of high-def replay.)

A major-league fielding coordinator can't just look at errors anymore to tell whether or not his shortstop is good. That was always a poor measure. A player might have great range, covering more territory, and at the extremes, have to make a hard play that might be called an error, where a lesser player might not even have a chance at the ball.

Derek Jeter, the Hall of Fame shortstop, is a dividing line for old-school and new-school thinking when it comes to defense. For his benefactors, they'll point to his jump throws, his dives into the stands, and his preternatural ability to anticipate a play, most notable on the highlight that's simply called "The Flip." Jeter, seemingly out of nowhere, comes into the frame as Jeremy Giambi is coming home. Jeter catches the offline throw home and quickly flips it to Jorge Posada, who tags Giambi, who didn't even slide. While we don't have the kind of metrics or even camera quality, highlights show that Jeter broke from covering second base at almost the moment of the throw, getting to the ball almost as it landed and making one of the most incredible plays you'll ever see.

Of course, the new-school guys will remind you that Jeter had terrible range, made spectacular plays because even routine plays were a struggle for him, and there's even an old joke that school children thought Jeter's first name was Pasta ... as in "past a diving Jeter," a not infrequent call when a ball was hit near the Yankees stalwart.

Both can be true. Just as Jeter was the last of the standard definition superstars, he also had near primitive tools to define him. Hall of Famer? Absolutely. Good shortstop? Well, I'll be charitable and say he was a great hitter for a shortstop. The fact is we didn't have the tools that we do now, nor do teams play defense the same way.

Today, players aren't even measured by a stat. Errors exist, but are generally ignored. Range factor? A relic. Players today are judged by predictive models that take positioning, expected speed, reaction time, and route to the play into account. From the time the ball is hit to the time it goes into the glove—or not—there's a percentage, a chance that the ball will be fielded. Think of a fielder as the center of a bullseye. Hit it right to him and a play should be made nearly 100 percent of the time. Hit it close and slow, and the percentage is high. Hit it farther out, or bouncing, or harder, and the percentage goes down more. At the extreme, a fielder has no chance at the ball.

Add in that there's seven fielders (plus the pitcher and catcher, of course, but their positions are more defined by the rules). Those bullseyes can overlap and of course, in the real world, they're hardly circles. Good fielders have bigger bullseyes, broader range where they're more likely to make a play.

"In a perfect world, I have fielders covering a significant percentage of an area. It's never perfect and if you hit it hard enough, it doesn't matter," said a current major-league fielding coordinator. (Because he wasn't authorized by his team to discuss how they handle fielding setups, even in the abstract, I had to grant him anonymity.) "Ask any third baseman. They're standing there 75 feet from a good hitter, maybe thinking bunt, and then the guy rockets one by him. You hear the ball but barely see it. If you make that play, it's luck. But again, why is he in that position against a guy who has the ability to turn on that pitch?"

That positioning becomes a combination of knowing the defender, the defense in a more holistic sense, and the hitter. Surprisingly, the pitcher doesn't factor in as much, according to the coordinator. "Most guys know what's coming. They can see the catcher, if not the signs. They know the situation, the batter, the pitcher's stuff, how's he's looked. Those things are in their head, but I need them to be in the best position possible. If they take a half step left because they see it's a change instead of a fastball, so be it."

Today's defense is predicated on Wee Willie Keeler—another Hall of Famer—and his famous advice to "hit 'em where they ain't." This is good advice, if you're as good a hitter as Keeler and not facing 100 mph fastballs. Hitters today can't aim the ball, but defenses can try to foil whatever they do by changing up where they start a play and making it more likely that hitters hit 'em where they are.

The shift isn't new. While many believe the first application was by Lou Boudreau's Cleveland club against Ted Williams, Joe Posnanski, the great sports writer, wrote about the Boudreau shift several years ago and quoted Hank DeBerry, an older scout, who said it had been done against Cy Williams when he played in the Baker Bowl (Philadelphia's "bandbox" ballpark) during the 1920s. Teams had made other shifts, bringing the outfielders in to prevent the winning run being one example, well back into the nineteenth century.

While the modern shift is a "love it or hate it" thing—and MLB seems to hate it, testing outlawing it during part of the 2021 minor-league season and proposing it for MLB—the fact is, it works. The better question is a naive one—why do we put the fielders where they are? How did this become the standard, and most importantly, is there a better way?

The setup has been the same way since even before the time of having nine men was standard. (The extras would be spread around the outfield.) Three men covered the bases, usually standing near it. The shortstop was so named because his job was essentially that of a cutoff man. He went out and caught the ball from the outfielder, especially if one had rolled a long way. Remember, no fences at this point and often high grass. Eventually, the infield shifted a bit, with players moving slightly away from their bags and the shortstop going into the area where most balls were hit. The first credited "standard" shortstop was Doc Adams, though more folks erroneously credit Dickey Pearce. Both men focused more on fielding rather than throwing.

What we know today is that each player might not be in the right spot. Given the tendencies of hitters and the known range of fielders in a 360-degree plane, there is a way to maximize coverage and computers can quickly determine it. This is just a more precise measure of how fielders have been positioned for years, by coaches waving them over, either by instinct or by scouting. In recent years, there have been so many shifts that one team used golf tees to put players in very specific positions. Naturally, the other team's fielder would just move them or pick them up. Today, they tend to memorize the spot, and this is often scouted by the team's own analysts checking to see the variance between where a player is supposed to be and where they actually were. The guys who can't find their spots had better not miss a ball!

Given this level of precision—Jones can range 70 feet to his left, Smith can range 60 to his right, so if they're 130 feet apart, almost nothing should fall—it's no wonder that players who were strict pull hitters were the easiest to adjust for. Bunching players into the areas of a spray chart that are populated by hits, overlapping the areas where balls have gone with fielders makes those "where they ain't areas" very small indeed. To put one in, it has to be hit very hard or, as David Ortiz likes to say, "they can't put fielders in the seats."

Bunt? A combination of ego and lack of skill tends to make that tough. Add in that it's hard to lay down a bunt when a pitcher is throwing 96 and spotting sliders on the paint.

Still, those types of dead pull, concentrated spray hitters are easy. They essentially cut the field in half already, so why waste a fielder or three in areas where, well, they ain't hitting 'em? But beyond the easy pull hitters, there are more subtle shifts that can help a pitcher. Teams have experimented with unusual-looking infield sets for guys that hit slow or fast grounders, or fast runners that necessitate quick throws—or throws that come quicker since the infielder is drawn in. Some have experimented with four outfielders or five infielders, adjusting based on data and situational need. While extreme, these types of shifts can work. With more subtle shifts, moving players a matter of feet rather than creating an extreme "all to one side" kind of shift takes away nearly as many hits.

A shift is more than just what Joey Gallo sees every time up. Defensive positioning is simply augmenting the defense's chance of keeping the ball under control in a game where the defense has so few options. While rule makers discuss banning extreme shifts or making four infielders have their feet on the dirt at the pitch, someone will figure out ways to adjust. One team I know has experimented at their complex league with what they call the "moving shift." Rather than staying stationary at the pitch, they put fielders in motion, staying with the "two on each side of second" rule, but quickly overloading the pull side in practice. The same kind of movement can be done in or out, or even pulling the left fielder in to have four "infielders" while the remaining outfielders stay in a typical "deep second base" pattern.

Don't expect rule changes to stop improvements in defensive positioning. In fact, new, creative solutions might create advantages for those that adjust the quickest, just like a good third baseman.

Framed

One of the dark arts of baseball is catcher framing. While men have been receiving the ball behind the plate for over a century, it wasn't until 2014 that anyone was able to quantify the value of it or show that one catcher was better than another at stealing strikes. This is of course a build-on of the technology to track pitches and one that interestingly is likely doomed by the same technology.

When PitchFX came in to existence in 2006, it took only a few years before Dan Turkenkopf was able to quantify the value of catcher framing. Now an analyst for the Milwaukee Brewers that has a great deal of influence on in-game tactics, Turkenkopf's work at Beyond the Box Score appeared to show that catcher framing had an outsized effect, able to create distinct advantages for pitchers, and that the act was a skill, repeatable by catchers despite pitcher and umpire, who should have a more direct outcome on the pitch. Instead, it appears that the pitcher just had to get it close and that the right catcher could create a strike by the way he caught the ball and how he made the umpire see that act.

As much as Turkenkopf's work, and that of those that followed with distillations and derivations of his original research, showed the value and skill of catcher framing, there was a mirror to that which is to this very day denied—that umpires can be fooled by catchers. To a man, umpires at every level say that how the catcher receives the ball or sets up doesn't matter.

In an interview, MLB umpire supervisor Jim McKean denied the existence of framing altogether, calling it "an entertainment word." Instead, McKean insisted that he and every umpire that he supervises is looking for one thing, the ball going through the strike zone. Of course, this zone is a theoretical construct. For some umpires, the zone is a cube, a three-dimensional structure that extends up from the four front corners of the plate from the low part of the zone to the high. (Ignore the rulebook definition of knees to letters. Almost no umpire does that, as shown by these same systems and by an infinite number of "K Zone" replays on TV.)

A more interesting recent version of catcher framing does involve the umpires, but they don't like it. Combining catcher framing skill—good and bad—with umpire bias—again, good and bad—appears to have a higher degree of real-world usefulness. Of course, umpires again deny there's any sort of bias, but it's clear that there are patterns, and even among the most accurate umpires in terms of calling balls and strikes, they do have weaknesses and even biases from umpire to umpire that are compounded by what's happening in front of them.

As I showed in a previous chapter, there are team analysts and consulting companies looking to find these little advantages and match them up. If a team knows a certain player is up and a certain pitcher has a pitch and a certain umpire is behind the plate and if hundreds more variables line up, the percentages can shift sharply into a team's favor. There have been binders of this information for years. Casey Stengel would famously reel off a story about even the most unique situations, almost always starting with "one time in the minors." However, even the mighty Casey, one of the greatest managers ever, is no match for a modern computer database.

All of which leads many to ask—why do we have the umpire at all? Joe Sheehan has written in the past that for years, the umpire's eyes were the best possible tool to judge balls and strikes. Today, that's not the case. We have 4K cameras in our phones. We have 8K TVs for sale at Best Buy. We have slow-motion instant replay from every angle and video rooms steps from every dugout. Why ask the ump to do something he's not the best equipped to do when the technology not only exists, but is already in place?

Umpires do a tough job and I'm not trying to send them to the unemployment line just yet. We're as far from not needing umps as we are from not needing warehouse workers and truck drivers. It could be closer than we think, but for now, we need them. Judgment calls and interpretations of the rules are better done by humans, for now, but the sheer number of calls—safe or out, is it a catch, fair or foul—that are better done by "the robots" should surprise many.

A robot umpire calling balls and strikes would, in theory, wipe out catcher framing and umpire bias, and make the strike zone much more consistent from game to game and even inning to inning. However, there would be another side. Some catchers have become very good at framing and it's a big part of their value. Taking that away with a stroke of the Commissioner's pen could take away a job.

I asked Xan Barksdale, one of the top catching instructors working with several MLB teams and USA Baseball, about this and he had a very interesting take. "I believe that the automated strike zone will have many changes to the catcher position. Some are known, more are unknown. Catcher may be the only position on the field where it's acceptable to have a defense-first type of player, and managers will gladly give up some offensive production to have an above-average defender behind the plate. However, I believe that may change when, not if, the automated strike zone is implemented."

Barksdale breaks down that the catcher has three main physical defensive responsibilities. "First, he has to receive the ball well. There are different philosophies about what that means, but most would say it's either 'keeping a strike a strike' or 'converting border-line pitches into called strikes.'" Both fit the definition of framing.

Beyond that, Barksdale says that blocking balls in the dirt and throwing out runners are the other two key defensive tasks. There are obviously more things any catcher has to do, but Barksdale emphasizes just how much more important these skills are to the modern catcher. "Yes, catchers have many other responsibilities like calling the game, fielding bunts, tagging plays at home plate, but these three are the Big Three for a reason!"

Even that third one may be devalued slightly and Barksdale agrees. "The throwing has become much less important due to the fact that stealing bases isn't a primary offensive strategy in today's game. Being able to receive the ball well is the most desired skill that any coach or scout would look for," he said.

That's shifting. "This has always been what coaches say, however, in reality, many coaches gravitate toward the catchers who throw really well because it's always been extremely hard to objectively evaluate how much impact a catcher's receiving skills have on the outcome of the game," said Barksdale. "With the advances in technology that have taken place over the past few years, we have been able to effectively score a catcher's ability to frame pitches. This has led to teams placing much more value on the receiving skill because it can now be measured more accurately."

Once that's taken away by a camera system that can't be influenced by a catcher's skill, that skill should drop immediately to zero value. Barksdale agrees. "The addition of a robo-ump will mean that a catcher's receiving skill, or lack thereof, will have no impact on the game because balls and strikes will be called using the automated system. I believe this is going to lead toward teams shifting from a defense-first mentality to them wanting to use a catcher who will provide more offensive production. This could potentially lead toward more offensive players being converted into catchers just to get their bat in the lineup."

Even the mechanics of the position might change, says Barksdale. "Another side effect will be that catchers' stances will change in order to optimize for blocking balls in the dirt. One of the latest trends has been for catchers transitioning into a one-knee-down stance in order to optimize for receiving and stealing strikes at the bottom of the strike zone."

Finally, Barksdale wonders how quickly this will make it into lower levels or even Little League, given the cost and complexity of this type of system. "It will also be interesting to see how this trickles down into amateur baseball and if teams who don't have automated zones follow suit and opt for a more offensive catcher over a defensive one."

I'm also curious if having the robot ump will change the mechanics of the position by changing where the umpire needs to stand. So far in experiments, the umpire is in the standard position. While he may not be calling the balls and strikes, he's still signaling them. There are other duties as well, such as judging tips and fouls, and calling fair or foul on balls down the line. Even something as simple as calling time is done better if the ump is closer or at least in the spot people are used to. For a catcher, it's unlikely to change much in terms of how he sets and fields the position.

Still, a move to a new system that looks to be more accurate and could change the game for the better overnight isn't going to be without growing pains and resistance. It also may change far more than balls and strikes.

Many a young budding General Manager grew up playing Strat-O-Matic, a card game of baseball stats and probabilities that likely led to many early sabermetric discoveries. Hal Richman's game was played by people like Bob Costas, Spike Lee, and Trip Hawkins, who founded EA, the people who bring us Madden football and more each year.

Dr. Rany Jazayerli, Joe Sheehan, and Clay Davenport found each other over a common love of "Strat" and of arguing baseball on Usenet, an early Internet message board. The three played in a play-by-email league, then met up with fellow player Christina Kahrl. They ended up founding Baseball Prospectus, alongside several others.

As far back as 2002, for his classic book *The Numbers Game*, Alan Schwarz asked how many GMs played Strat as a kid and found that almost half of current GMs had played it, coloring the game in the early post-Moneyball era, leading in directions where we may only now be seeing the full impact.

Dan Okrent was a Strat addict, but he shifted his obsession, instead developing a simplified version that used real stats. He and several friends went over the notes he had jotted on a flight home while at the New York restaurant La Rotisserie Francaise. The rules they put together came to be known as Rotisserie Baseball and led to an explosion of fantasy baseball leagues, especially as the Internet made it possible to get stats quickly and automatically. Okrent was at the time working for *Texas Monthly*, but he went on to do some other things like being the editor of *Time, Esquire, Life*, and was the first public editor of the *New York Times*.

Okrent also wrote a profile for *Sports Illustrated* of an unknown baseball analyst named Bill James in 1981.

This is a long hagiography of a card game, but one of Strat-O-Matic's lesser-known effects was the concept of the defensive spectrum. James was the first to name it, but any Strat player knows the concept—certain positions are more difficult defensively, and moving down from the hardest to easiest creates a loss of value. This was later quantified in many ways, but James's work was almost pure theory at the time due to a dearth of good defensive statistics. There was no dWAR, no StatCast, and even James himself hadn't developed Range Factor.

The basic theory is that the spectrum goes from left to right in terms of difficulty:

Shortstop - second base - center field - third base - right field - left field - first base

As a player can't handle a position, he's shifted down/to the right. The positions of pitcher and catcher were listed by James as to the left of shortstop, but later iterations leave them off entirely since the skill set is so different. The same is true of designated hitter, the ultimate "move off the spectrum" placement.

In 2021, many teams were willing to shift players up and down the defensive spectrum. The Mets traded for the Cubs' shortstop, Javier Baez, and shifted him to second, a standard move. The Rays traded for longtime DH Nelson Cruz, who had only played nine games in the field since 2016. This isn't new, with players moving down the spectrum for years—Ryan Braun's bat had to play, but his glove couldn't at third base. When Alex Rodriguez came to the Yankees, he was the better defensive shortstop, but the team wouldn't shift their captain, Derek Jeter.

The spectrum isn't scientific, or even accurate. A number of players, especially in recent years, have shifted from third base to second with success. Large players like Cal Ripken Jr. were regularly tagged for a shift, but Ripken's early success allowed him and others behind him to stick at the position as norms were challenged.

But what if Bill James were wrong? Today's offensive environment and focus on putting the ball in the air makes traditional defenses execute standard plays less and less. Shifts can move players around and are often made by "everyone move left a bit" rather than putting the rangiest defenders in an area covering the most territory. If a third baseman can shift to play the shortstop slot in a shift, with no help to his right, why couldn't he play a passable defense in a standard set?

The question comes down to defensive value, which is still questionably valued. In 2019, the last full season available, the WAR leader for the major leagues—in theory, the most valuable or best player—was Alex Bregman of the Houston Astros at 8.9 wins. This was largely due to the defensive value the system saw in his play at short. There were two more shortstops in the top five, showing just how much defensive value is implied at the position. In 2018, when Bregman played only 18 games at short, his defensive value was calculated at half a win lower.

Most players shift down the spectrum because they prove they can't play the position, but what if we shifted everyone up—to the more difficult positions—and saw what happened? Most recent results have been minor and changes in the game make it easier to cover for deficiencies. Imagine a fly-ball pitcher playing against a fly-ball lineup. Why have a good fielding infield if there's a hitting advantage to be had over those reduced chances?

This is an ideal situation. Most teams barely have a couple extra bats on the bench, let alone a squad that can platoon out at something esoteric, but it is possible in theory. Whether that theory is ever tested or not remains to be seen, but new technologies and new training techniques are giving us players that are bigger, stronger, faster, and way more athletic. I mean, how bad could Aaron Judge be at shortstop?

Fielding is perhaps the most progressive area in baseball, but there hasn't really been innovation besides re-discovering the shift in recent years. Talk that the shift might be banned make even that tougher. With more and more data and better athletes, someone's going to figure out a better way to play defense.

Though I suppose, maybe Wee Willie can be proved wrong. Maybe there's a defense in the future where there's no "ain't."

7

BASERUNNING

There is a simple base-running formula. Trevor Forde, a coach at the University of Indianapolis, explains it in the following manner:

A - the pitcher begins his motion to the plate and delivers the ball.

B - the ball is released and gets home, hitting the catcher's glove.

C - the catcher receives the ball, transfers to the throwing hand, and throws.

D - the ball is thrown to second base and goes to the infielder's glove.

E - the infielder applies the tag.

Now, at the same time, the runner is putting his side of the formula together.

X - runner takes his lead, shortening the 90 feet between first and second.

Y - the runner accelerates, runs a portion of the 90 feet, and slides.

Z - sliding, he touches the base, but he does not slide past it enough to lose contact.

There are known times for each of these, in terms of averages, which help us frame the formula in more concrete terms. The pitcher takes between 1.3 and 1.5 seconds, on average, to deliver the ball when there's a runner on base. A 90 mph fastball takes about 0.4 seconds to get from hand to mitt. The catcher's "pop time," or time from mitt to throw to the man covering second base, is between 1.9 and 2.2 seconds. It takes between 0.2 and 0.4 seconds for the infielder to apply the tag, ignoring evasion or other quirks.

At best, the pitcher to tag (A through E in our formula) is just shy of four seconds.

On the other side, everything is a bit murkier. Lead lengths vary. Reaction times vary. Speed and acceleration are measurable, but they vary too. The slide—again, just the standard version—depends on length, which part of the body goes first (head or feet), and the surface.

The simple part is that a runner knows that if he can do what he needs to do in four seconds, then the other side—pitcher, catcher, and fielder—have to do things perfectly and without error to get him. As Harry Callahan once said, "Do you feel lucky, punk?" Runners shouldn't feel lucky; they should know, and that's the science of modern baserunning.

One will be faster than the other. One will make fewer errors or execute better, or worse, and in the end, we'll know which was right. Four seconds, or thereabouts, that can change a game. Go ahead, count that out and think about all that has to happen and all that could go wrong, in about the time it took for you to read this sentence.

For Forde, this isn't a guess. It's a probability.

Forde's key is trust. He believes that all of this means, for a baserunner, the chance of success is there, but not without trust. "My guy has to believe it, but he also has to trust me, trust the coaches," he explained. "We're not going to punish him because he gets thrown out. It happens. If the pitcher makes the throw, if the catcher beats his pop time, if he throws on line, and the tag is down, and my guy's out, that's just baseball."

Forde is often the first-base coach during games. It's a thankless job, down there in the box in a hard hat, swinging the ever-present stopwatch. Some innings, the first-base coach can look like the equipment manager, taking gloves and guards off the guys who get to first or even second base. But there's far more going on there.

I asked Forde what he's looking for when he's out there. He's already spent hours going over scouting reports, watching video, but the in-game reads are what look like nothing, but are clearly key. "I'm looking for tells," he said. "It can be something like breathing, timing, anything. A lot of guys give you one look, two looks. Most of them aren't even looking over, really. It's maybe peripheral vision."

Indeed, many teams have moved to a different model, where the third baseman will give a small signal to make a pickoff throw. It's a wink, a nod, a glove movement, so that a runner who's thinking that the pitcher isn't paying attention to him shifts a little bit and finds himself out too far.

"Breathing is one of the things I'm looking for," Forde explained. "Especially at lower levels, these guys have their heart rates up, they just let a guy on base, now they're trying to stay out of a big inning. They're pausing more. They're thinking more. A lot of times the guy will just take that breath and try to get back in his zone and that's the spot where he's forgotten about the runner."

Forde, like most, realizes that speed isn't everything. It helps, but there are plenty of fast runners who aren't good baserunners. "Sometimes that 6.5 guy doesn't play on the bases," he said, referencing a 6.5-second 60-yard dash speed, often used as a measure for baseball speed. "For most guys, it's about the lead. They get out, they're keeping themselves centered. They know they can get back, they're not leaning too much."

But runners do get picked off, even on Forde's watch. What happens? "Typically, if you get picked off there, it's more of a mechanical thing," he explained. "We always teach that you keep your feet underneath you. As soon as your feet get outside the framework of your body, it's going to be tough to overcome that. We're not necessarily trying to cheat, we're just trying to put ourselves in a better position to move toward that base. A guy makes a big movement and then he's no longer within the framework of his body. His lateral movement is trapped, he's frozen, he's out."

Leads—the ability of a baserunner to get off of first base, cutting the distance from 90 feet to somewhere between 70 and 80, changes the formula. Even with an imperfect reaction or jump, or with less-than-ideal speed, the time to get from first to second is reduced.

Consequently, more pressure is put on the defense and that pressure can lead to mistakes. Forde knows this and measures it. "Even with a jump lead (where a player hops out and back, attempting to get a little more distance or momentum to second on his break), the runner has to be able to get back," he explains, standing as if he's on base himself. You can see him envisioning it. "A 12-foot lead is standard, but if you're stretching to 15, you can't do a jump lead. You shorten the lead for the jump so you can still get back, but you have to keep your feet under you until you know you're going. You have to be as committed to the return as you are to going."

Just the threat of stealing alters things and gets the defense thinking, changing, worrying. Forde wants to use that threat to force mistakes, not only in the base-running phase but in general. "If there's a good lead, a fair lead, it's almost a negotiation. I go to here, the pitcher doesn't throw but if I go just a bit farther, he does," Forde explains. "Even that throw is a potential positive for me. There's too many times I see that guy give up the throw, lets the first baseman get the ball, easy. The pitcher is already uncomfortable. He doesn't want to throw to anywhere other than home. That's what he practices. If a team works on picks, it's usually an afterthought, a few times maybe once a week and usually it's more for the runners. I want my guys to square up, hold their position, and if the pitcher throws it away, I don't have to steal the base, it's mine and maybe third. That's two bases just by a threat of stealing and the right footwork."

Forde knows that it's not just the throws to first that change the pitcher. "My guy's on base and the pitcher's outside of the comfort zone. He's not thinking as much about the hitter, he's thinking *I don't want to throw it away*. He's thinking *I don't want to bounce one in*, so it's more fastballs. It's more slide steps so he's losing a tick off the fastball he might not want to have thrown in that situation. It's a circle. I'm creating offense by dictating to the defense. I'm creating runners by focusing on on-base percentage, then stealing slugging by in essence creating a second bag. A weak single or a walk? That's a double now, so I don't need a real double to score, I just need another single. I'm creating chaos."

At the collegiate level where Forde excels, there's more of an opportunity for mistakes. "I see missed throws every game," he said. "It's part of it. JT Realmuto doesn't sail the throw [meaning throw it high, into center field] but guys at [this level] do. That factors into the percentage, but I'm still looking for 80 percent. If I tell you you're going to win a bet 80 percent of the time, you're taking it, right? And you're not crying about the 20 percent that are losses."

I'd heard this line of thinking before and disagree with it. I responded that while I could, in theory, bet on every NFL game in a given weekend, which is between 14 and 16 games for 18 consecutive weeks, where an 80 percent win percentage would be shockingly high, that's not the case in baseball, where a game is 27 outs and those don't carry over. Running at an 80 percent clip is going to be good on a seasonal basis, but in a game, the situational math doesn't always work out. It's a sample size of one, always a new situation each time.

The Value of a Steal

While basestealing has long been considered a big part of the game, it's a little surprising when you learn, as I did in researching this chapter, that caught stealing wasn't even kept as an official stat until 1951. Even how a steal was defined wasn't locked into the modern interpretation for the first quarter century of the game (1898).

It ebbed and flowed with both speed and power. Ty Cobb stole bases, but Babe Ruth knocked them in with his giant bat. The game's speed was way down in the 1940s and 1950s—Dom DiMaggio led the league with just 15 steals in 1950. As pitchers became more dominant and Astroturf sprouted up around the league, steals became a better deal for runners and they led to a renaissance of the art, with players from Lou Brock to Vince Coleman redefining it.

This chart, put together by my research assistant David Barshop, shows the ebb and flow of steals percentage, but mostly that the multi-tiered expansion of the game really changed things:

1951	.589 (total caught stealing: 604) (total steals: 866) – 16 Major League Teams
1961	.642 (total caught stealing: 582) (total steals: 1,046) – 18 Major League Teams
1971	.629 (total caught stealing: 1,039) (total steals: 1,765) – 24 Major League Teams
1982*	.662 (total caught stealing: 1,617) (total steals: 3,176) – 26 Major League Teams
1991	.678 (total caught stealing: 1,475) (total steals: 3,120) – 26 Major League Teams
2001	.687 (total caught stealing: 1,408) (total steals: 3,103) – 30 Major League Teams
2011	.722 (total caught stealing: 1,261) (total steals: 3,279) – 30 Major League Teams
2021	.755 (total caught stealing: 547) (total steals: 1,689) – 30 Major League Teams

*1982 was used rather than the sequence of "1" years because of the strike in 1981, which reduced the number of games that season and would cause comparability issues.

Surprisingly, the immediate "Moneyball" era where analytics argued against the steal took a while to catch up. Almost a full decade later, steal attempts were about static with a much higher success rate. The drop in the following decade is where it really locked in, as power and pitching made losing outs on the base paths even more costly. Steals did not decline as much as most thought in the "steroid era," but dropped quickly as analytics took hold in the 2010s.

The question is whether the same kind of Moneyball/arbitrage advantage could be found in speed. The NFL has placed a premium on speed over the last two decades as they've shifted from a power game to a passing game. Here's a thought for one of those tanking teams—bring in a bunch of speed and see what happens to both defense and base running. If sabermetric orthodoxy holds, they'd still tank, but at least they'd be entertaining.

In today's game, there's no Rickey Henderson, but Henderson is one of the greatest players ever. Bill James said that if you cut Henderson in half, you'd have two Hall of Famers. Of course, it's folly to try to will another Rickey Henderson into existence. A journeyman who is as quotable as Yogi Berra, Henderson blew away almost all the stolen-base records, finishing with nearly 500 more than Brock's previous record.

Henderson started in a steals and small ball era, but also had power, with almost 300 home runs in his career, but by the end of Rickey's run, there were power speed guys like Jose Canseco hitting the "40-40 Club," 40 homers and 40 steals in the same season. Only four players ever made it and three of them are under a cloud of steroid use. Henderson's season high for homers was only 28, but he stole 66 bases as late as his age-39 season. Power ages well, but speed doesn't.

No less an authority than Rickey Henderson would regularly say it's easier to steal third than it is second. The pitcher doesn't have a clear view, the lead usually is a bit longer, and with a right-handed batter, it's a tougher throw, if shorter, for the catcher. The formula is still the same and for Henderson, his success rate was slightly higher—82 percent versus 81 percent. Henderson attempted to steal third much less often—about 400 attempts at third versus over 1,300 attempts at second. I couldn't find a good count of how many times he was on second with a realistic chance of stealing, since so many factors go into it. It's not as simple as "How many times was Henderson standing on second with third open?"

However, Forde puts some conditions on agreeing with Henderson. "At the big-league level, you don't have some guy shading in to second. You're just not going to open up that big a hole for the hitter, so you let the guy have another step. There's also all sorts of metrics and tendencies and you know which guys might go. Maybe you tighten up on ... who's the top steal guy now? Trea Turner? [Whit] Merrifield? There's just so little running at the major-league level that I don't think we even really see much of an adjustment. Everything's perfect."

It's different at lower levels, but Forde still doesn't push for stealing third. "It's really matchup. At second, most of the runners can score on a single, so I think the percentage has to go up."

One thing that Forde thinks is underrated is the ability to slide. I asked him if he felt sliding was a skill or a talent, with the difference being that a skill is something you can practice and get better at, like hitting, or a talent, like speed.

"I think it's talent. It's tough to teach," Forde said. "Athleticism obviously plays a major factor into that slide. If you're taking a 7.2 (60 speed, below average) guy that has great baseball instincts, he's not gonna probably be the best slider. Get a 6.7, a 6.5 even, and he just naturally has more athleticism and seems to avoid tags. There's a feel you gain, so that's a skill. The more you do it, the better feel you have. You have to just get comfortable doing what you're doing to be successful. But sliding is definitely a talent. I think it's just hard to teach. We don't practice it. I've never met a coach that said, 'Hey, we're gonna go do slide drills, 100 slides into second base.' The players would hate you, and you're beating them up over the course of a long season anyway."

The baserunner is obviously key, but what about the coach? The first-base coach almost seems like an appendix, a guy who's standing there with a stopwatch and collecting shinguards as much as acting as a part of the game. Forde laughs at the image, but thinks most people don't see what the first-base coach is actually doing. "I'm trying to understand the whole situation, starting with who I have on first base. There's things I key off. I want to know how my guy's return is, right? Can we get another half step? Is that where we normally stand, or is he conservative? I want that at the ragged edge."

Forde continues. "Then I'm looking at the pick. Is it under a second? I see 0.9 and that's good. That draws my attention. If he's over a one, I'm telling my guy to get another half step. I know the pop time. After that, it's all just sequencing. If you don't get a good jump, stop and come back. It's not like you have to go."

At third base, there's less concern with stealing and more with the finer points of baserunning. The third-base coach in the mind's eye is up the line, watching the play and either waving his arm wildly to go home or throwing up a stop sign where you can see him willing the player to a standstill. In a role where only the extremes are noticed, positively and negatively, I asked Forde if he thought that made coaches more risk averse.

He didn't mince words. "I think they're timid. Don't get me wrong, I think they do an unbelievable job. I was watching the Yankees–Red Sox game today and I thought the guy was going home and the third-base coach held him up. It's Yankee Stadium, so short right field and just as I'm thinking he should have sent him, the throw is on a line, perfect to home plate and I realize the coach probably knew that. He'd seen the video. I looked it up and Hunter Renfroe, the Sox' right fielder, has 14 outfield assists. That's a lot. He's normally a center fielder, but the Sox have him in right and he's playing in a short field. That's a lot of information, plus you have things happening in front of you, so good on that coach!"

One lesser aspect of the base coach is the interrelationship with the first-base coach. "It's key to be on the same page, same wavelength," Forde explained. "If the third-base coach is sending someone, I have to look at the play at first. Do I send that runner to second right off or is he holding so he doesn't get the shorter throw to get him, especially with two outs? There's subtle things like, do we have a right-handed center fielder or left? If you know that, and he's going so he has to throw against his body or make a turn, that's a bit of extra time and more chance for him to make a mistake."

Forde thinks another valuable thing is just keeping a player in the game. As simple as it sounds, saying something like the number of outs can be important. "Think how many times you watch a game and lose track of how many outs so you look at the screen," he said. "Especially with kids, they get their heart rate up, they just had a big hit, and they'll just lose track. You see that at every level. Even the umps have their signals so they don't lose track. I do the same things. I want to keep the verbiage simple—two outs, Vaughn's speedy at third. If we've got a play on, I can't give it away. In the play, it's simpler—tag on this, but the runner has to have been listening, that we both know the center fielder's a lefty and going away from it, so tag. There's just so much information but so few words."

At the end of my conversation with Forde, I realized I'd probably never had as long a conversation with anyone about baserunning. I looked for books on baserunning and while there are tons of them, they tend to be very simplistic. There's never really been a classic book like Ted Williams's *The Science of Hitting* for baserunning. It's not because it's easy, or not important, but that it's never really been broken down and put into form, especially as stealing bases has lost importance over the last couple decades.

If Rickey Henderson isn't going to write the book, maybe Trevor Forde has a new project.

8

TRAINING AND DEVELOPMENT

There's a famous story about Nolan Ryan. The sportswriters came in to interview him and were forced to wait. As he walked out of the weight room, uniform still on and dripping sweat, long after the game, the writers realized he'd been lifting hard after throwing more than 100 pitches.

Ryan was unusual in a lot of ways, but he was also one of the pitchers that wasn't afraid to go in and lift some weights. He'd been strong since a young age, growing up on a farm, but most pitchers steered clear of the weight room—as did most players—worried they would lose flexibility if they bulked up.

While heavy weight lifting is now accepted and even encouraged, there's no single accepted development path in baseball. Study after study has shown long toss to be safe and effective, but there are still major-league teams that limit long toss to 120 feet.

Why 120 feet? That comes down to Dr. Frank Jobe, the doctor that also created Tommy John surgery. Jobe told me in 2013 that it came down to the distance that was available at Dodger Stadium for rehabs. Once a pitcher got to that point, Jobe would release them back to the team. It was never intended as any sort of limit, but the number somehow got memorialized in rehab protocols and the institutional memory of far too many pitching coaches, even fifty years after something occurred randomly.

I'm telling you, baseball is slow to change.

One area where baseball does accelerate the pace is in technology. We've seen more advancements in measurement and data over the last five years than we have in the hundred-plus behind it. From the deployment of Statcast and Hawk-Eye, we've seen technologies and tools rapidly come down to the youth levels.

There's probably not a top facility that doesn't have a collection of tools like Rapsodo, HitTrax, ProPlayAI, and more. Video analysis is everywhere, as is gamification.

"When we brought in HitTrax, I knew the kids liked it," said Al Ready, the head coach at the University of Indianapolis. "I wasn't expecting kids to take to it like a video game. One of my players has thousands of swings, all analyzed in there. He's trying to hit the ball out of the park in San Diego like Tatis and I'm watching his bat speed go up and up. We have challenges where we go for top bat speed, farthest hit, things like that. And when that kid I mentioned took a ball out for a walk-off homer in the conference tourney, maybe he wasn't thinking about all that time on HitTrax, but I sure was."

The downside is that HitTrax isn't cheap. It's a multi-thousand-dollar system that requires maintenance and inevitable upgrades. However, there's a new run of systems that are using the power of the supercomputers we all carry around in our pockets. One such system is ProPlayAI, which bills itself as a portable biomechanics lab. The accuracy isn't the same as a million-dollar purpose-built lab, but 95 percent of the value at less than 1 percent of the cost is one of those things that Apple and Google have been telling us was coming. Moreover, since ProPlayAI works from a single phone, it can be set up anywhere, not just in a dedicated biomechanics lab. A similar app, Mustard, basically puts renowned pitching guru Tom House into an app.

An app called HomeCourt was demonstrated at the most recent Apple iPad introduction. It uses the iPad's camera and integrated laser rangefinder to become a digital umpire for tennis. Similar systems are under development for baseball, and going from multi-camera setups that cost in the hundreds of thousands to a couple iPads is a huge step in getting better technology to the masses.

There seems to be no place that's off limits for tech and development. Sleep can be tracked by an Apple Watch, an Oura ring, or specialized devices, even some built in to beds. Heart rate, heart rate variability, and even something like VO2max (the maximum throughput of oxygen a person can hold for a period of time, an oft-used measure of cardiovascular health) can be calculated and put into a training program.

With all this data available, it's no wonder there's a secondary market of apps and services helping people understand all the data they've been collecting. Coaches aren't biomechanists, statisticians, or physiologists; expecting them to be seems a bit much, but coaches have either picked up enough to get by or have been left by the side. Those that can learn and integrate the rest of the data into a program are going to have huge advantages. Those expensive certifications you see are just one mark on a baseball coach's resume these days, but they are noted.

Carrying the Weight

One of the biggest controversies of recent years has been over the use of weighted balls. Weighted balls have been around for years, but they were really popularized by Kyle Boddy and his Driveline Baseball outfit over the last five years. Boddy's pitching development facility has become one of the hotbeds of research and something of a mecca for pitchers at all levels. I've known Kyle for years, since 2005 when he flew to spend an afternoon learning the gyroball from me at a local facility. (Boddy and his crew have come a long way since then.)

One of Boddy's focuses has been weighted balls. In a variation, Boddy also created soft but heavy balls called "plyo balls," which is a variant on plyometric, a type of exercise. The use of both types of weighted balls has been key to Driveline training for years, but I do want to enter a caveat here. Driveline training is extremely individualized and often hands-on. It is easy to make changes there and their staff is highly trained as well. Those using what they think of as Driveline training—but using only part of the program or using it as a cookie-cutter approach—is where many of the issues come up.

Weighted balls can and do help many, but they also can cause injuries, as multiple studies have shown. The difference is how they are used and how well prepared the athlete is to use them. It's not unlike a chain saw—a lumberjack can cut a lot of trees with a chain saw, and an idiot who doesn't understand what he's doing can cut his leg off. Both are true. The issue is how the tool is utilized.

Weighted balls, in and of themselves, are no more a tool than any of the other innovations that Boddy has helped to popularize. "Run and Gun" throwing, where a pitcher runs forward, crow hops, and throws the ball at high velocity, has been around for years. Long toss has long been the realm of Alan Jaeger, the prototypical California peace, love, and pitching guru that helped Barry Zito a decade prior. Dr. Mike Marshall, the Cy Young winner and pitching iconoclast that passed away while I was writing this book, also advocated weighted balls, but his were often really weighted, like shot puts.

Instead, Boddy and his Driveline crew were able to put all of this together and popularize it. They were able to get notice, where others were simply niche guys, who never got the reach that Boddy did. Moreover, Boddy produced results. Pitchers at his facility showed velocity gains of the type that normally come in the back pages of *Baseball America* and are about as shady as back pages everywhere. "GAIN 5 TO 10 MPH IN WEEKS!" Well, Driveline guys did just that and backed it with science and data.

Success breeds success, and imitators. The problem is that imitators didn't notice that Driveline was doing individualized training for largely elite-level pitchers. That meant the imitators started handing heavy balls to everyone and saying, go at it, kid.

This was like handing a chainsaw to someone with no training. It sure can cut down some trees but in the wrong hands, you end up with a lot of injuries, and in this case, weighted balls could seem just as dangerous to limbs.

Studies showed the problem—that putting younger, undeveloped, and under-prepared athletes into a weighted-ball program could create arm issues. The same people that didn't read the first time around continued not to read. Sentiment quickly whipped from "weighted balls for every one!" to "weighted balls for no one" when the truth was something completely different, but right there for everyone to see, if only they'd, you know, read.

Ironic that I have to say that in a book.

The fact is that weighted balls are just a tool, but an effective one for a pitcher that is prepared to use them and that has the right kind of instructor that can adjust as needed, monitor closely, and tweak the rest of their development to make the best use of the weighted balls and any other tools.

Anything can be abused. Weighted balls shouldn't be seen as anything more or anything less than a tool, albeit a potentially powerful one, in both the positive and negative. As with anything in baseball, proper coaching is the key.

Most of you reading this likely played baseball up to some point. You played tee ball, then coach-pitch. You moved to the Little League, then the big field. You tried out for the high school team or if you're younger, the travel team in your area. Maybe you went on to college ball or even the minors. If we follow this pyramidal system of baseball, there's probably one or two of you that actually played in the major leagues. (Hi, Cody.)

You probably remember those coaches you had, the ups and downs, and the connection you made with your coach.

In America, those coaches are largely dads. When I coached a team of 16s a couple decades ago, people would always ask, "Which one is yours?" Dad-coaches aren't inherently a bad thing, but the lack of training or even basic education is problematic and far beyond the fact that I simply can't hit fungoes. Believe me, I've tried.

Go to almost any youth league in any sport and you'll find that at best, you might have to take a first-aid course and maybe a one-day course that barely covers the basics. This isn't to say that some youth coaches aren't good or that they don't take it upon themselves to get additional education, to teach themselves to get better, but it's certainly not mandated. Between costs and limits on practice times, even professional education for upper-level coaches—high school and college—is few and far between. Athletic trainers and physical therapists go through thousands of hours of practical education, while the coaches they work alongside—and too often, under—have none of that, at least formally. The support of these coaches is simply lacking.

This is an absolute failure of Baseball—capital B—as a sport in America, which has largely been due to a hands-off approach from the Lords of Baseball. Europe has a system of licenses (licences, that is) for all their football coaches. Canada has a set of regulations for coaching, with prescribed courses and hours of practical education. America requires less training to be a youth coach than it does to get a driver's license.

Baseball in 2020 started a program that Rob Manfred introduced as "One Baseball." In it, he said that baseball needed to control not just the major leagues, but the minors. They needed special rules that would exempt baseball teams from minimum-wage rules, or else teams would need to be contracted. They got the special rules, passing them through Congress, and then quickly contracted the teams they said they wanted to save.

By 2021, the entire organization of minor-league baseball had been overturned and MLB had become capital-B Baseball. The next step is to take over collegiate baseball. NIL rulings and a structural reduction of the NCAA is going to accelerate this, but offering a two-track system—the NCAA is fine, but why not be in our Draft Combine as well?—put MLB into a stronger position in 2021.

Baseball development is in a weird place. It's stronger than it's ever been, while making baseball almost an exclusionary sport. It costs thousands of dollars to join an elite travel team and also forces single-sport specialization, yet the best players and best coaches are gravitating to this because that's where the best players and best coaches are. It's a vicious cycle.

The creation point of travel baseball was far in the past, but the modern version of it accelerated largely because of pitch counts. When Little League put in pitch counts for youth players, there was a quick abandonment of it. A rule designed to help arms forced a structure that tears them down in all too many cases.

This isn't to say that all travel baseball is bad. It's not. It is, however, expensive, exclusionary, and reductionist. What baseball has to do is combat the parts of a system that exist, filling in the holes with enough incentives to not leave another generation of urban and rural kids behind. While baseball players are considered costs, not assets, losing baseball at lower levels—or even reducing it with reduced numbers, poor training techniques, or increasing injuries to the best of the best—makes the product worse.

That's not development. That's destruction.

Getting Progressive

In 2006, I took a look at the landscape of pitcher training and pitcher injuries and wrote a piece on what I believed then to be the best possible system for developing pitchers at the major-league level. It's sixteen years since then and I still believe that this progressive development system is the best hope of building better pitchers. I still don't know if it's foolproof, because no major-league team has tried it, despite its simplicity, basis in science, and quick and easy read on results. What I can tell you is that this system wouldn't hurt anyone and if you get a couple months in and it's not working, there's not much you have to change, nor is there much real change to get into it.

Since the dawn of the free agency era when pitching became scarcer and more expensive, teams have stopped treating pitchers like a fungible, replaceable commodity. Unfortunately, their strategies for keeping their pitchers straddling that mythical line between healthy and effective have been, at best, hit or miss. In fact, there's almost no science in pitcher development. The training methods of Johnny Sain and Leo Mazzone are simple folk wisdom passed down from guru to student, with the student eventually becoming the new guru. Of course, this wisdom does stand up to many of the tests science throws at it. With newer pitching coaches, science is entering the picture with Tom House, Rick Peterson, and Glenn Fleisig leading the way. Still, we're leaving something behind and seeing far too many pitchers headed to the operating table.

The work of Craig Wright, Keith Woolner, and Dr. Rany Jazayerli on Pitcher Abuse Points, as well as some subsequent work by others, has shown us that there is certainly some value in pitch counts. However, pitch counts and PAP work in the average—what describes pitchers as a whole does not necessarily describe each and every pitcher. In fact, it is this variability that is often cited as a criticism of the system (the inability to describe outliers was notably debated in *The Neyer/James Guide to Pitchers* by Bill James, Woolner, and Jazayerli). Apart from pitchers like Livan Hernandez—who still pitch in a Mathewsonian style—where they "coast" and save their best stuff for when they need it—there are pitchers who seem to be able to take a higher-than-average workload over an extended period of time without discernible or predicted negative consequences. It was three years ago when I predicted that C.C. Sabathia would break down—and I'm still waiting.

Still, there is a value in understanding which pitchers can exceed the averages and become the outliers, maximizing their talent and usability. Instead of merely understanding how pitchers become injured or ineffective, is it possible to use the science of PAP and marry it to the coaching-centric current system? That's the motivation behind the progressive development system.

Progressive development is a very simple concept that takes the tenets and rules of PAP and molds it to a system that would work with very little adjustment inside the current, change-resistant baseball system. Progressive development is already an accepted principle of weight training and physical therapy. In fact, the minor-league promotion system is itself a progressive development system. One wonders why this approach has never been used for pitching or if in fact it has, only quietly. Parts of the system have been in place, but never in an organized and complete manner. The simplest way to explain progressive development is to put it in terms of weight training. No one bench-presses 300 pounds on their first day. They work at it, gradually adding weight to the bar and, over time, grow stronger and stronger. Some will be stronger than others and each will have an individual maximum that will fluctuate slightly.

The key question becomes: How is it that pitchers stay healthy in the absence of low pitch counts and Mathewsonian technique? The key may well be the progressive approach that was taken. Often, this approach was taken by accident and the pitcher himself is known at a young age as a workhorse, able to take on workloads that break down lesser, weaker athletes. If we accept the Sain-Mazzone tenet of "throw more, pitch less," then we are also accepting the principles of progressive resistance. Throwing is the equivalent of strength training, one five-ounce lift at a time.

A more scientific and less random approach would be better, and there are simple adjustments to the normal development curve used by major-league teams that could employ this. It's important to note that the approach should work at any level, including the high school or college level and, at least in theory, this should be the beginning of any enlightened approach. In practice, high schools and colleges have turned into arm-shredding machines, forcing pitchers to take on workloads that approach the criminal, with the survivor effect the most noticeable benefit of drafting collegiate pitchers. The current approach will begin with the assumption that the pitcher is being drafted into professional baseball, where the initial low-workload stages in the low minors serve as a quieting, calming period that should negate some of the short-term effects of overwork.

A pitcher is initially drafted and assigned to rookie ball and should have a hard pitch-count limit based on his previous workload. These short-season leagues serve as an important adjustment period and, in fact, seem to be very effective for just this purpose. Extremely low pitch counts, as low as 60, should be in place. This serves multiple purposes. First, it is again a calming period for the pitchers who have been previously overworked. Second, the low count should serve to protect the arms of young pitchers with mechanical shortcomings. (Surprisingly, some of the more enlightened teams do no mechanical work with their draftees in the first short season by policy. "I want to see them pitch. The work comes in the offseason," said one minor-league pitching coordinator.) Finally, the low count forces the use of multiple pitchers, allowing the largest number of draftees to have a chance to develop while also putting these pitchers in various roles. Since almost all pitchers will be starters at this stage, allowing them to start games, close games, and pitch in relief will help give them a taste of what they may one day be asked to do.

At the single-A levels, pitchers will begin the season on a minimal number of pitches, again beginning as low as 60. The more advanced pitchers will be paired in the tandem starter system, alternating as the starter and the reliever. Any relievers will also be on pitch counts, though they will likely not be reaching them due to their reduced role. So far, this is no different than the tandem system already used by several teams with excellent results.

Where the progressive system differs is in how the pitchers are allowed to develop. Simply put, pitch limits are set by three factors: results, goals, and health. Pitchers are allowed to increase their in-force pitch limit by five or ten pitches each time they complete three starts in an effective manner, come back for their next start without limitation, and complete goals for factors that the team sets (such as first strikes, ground ball ratio, or more individual goals, like mechanical changes). Even with the increases, there should be an emphasis on pitch efficiency. There is no reason that any successful pitcher cannot complete five innings—enough for the win—on 60 pitches.

Once those goals are reached in a three-game "set" (three consecutive games), the pitcher is allowed five or ten more pitches per outing. This will allow a pitcher to progress throughout the season, a satisfying thing for a young pitcher looking to succeed while protecting his arm in both the short and long term. If a player goes from Low-A to High-A or repeats a level, the pitch count should be set to a number somewhat lower than the ending count for the previous season. By the time a player reaches 85 to 90 pitches, he can safely be removed from the strict tandem system. Any pitcher that cannot work to these levels can be shifted to a reliever track or washed out of the system.

At the double-A level, pitchers often face their toughest tests. It's no different in the progressive system. Pitchers again start the level with a slightly lower limit than they ended the previous season with and are allowed an increase after a five-game set of successful outings. The pitcher is proving that he can accept more workload, and that he knows the specific things he will need to do to advance. Since it is implicit in the system that a pitcher will be at 80 pitches or more at this level, efficiency can be introduced, essentially asking the pitcher to complete more innings on fewer pitches. Ideally, the pitchers at this level will be in a four-man rotation, teaching them to recover from starts quickly and, since single-game pitch counts should be relatively low, the recovery should not be a problem. Pitches should be capped at this level at 120 for pitchers under 25. Research shows that pitchers have not only passed the "injury nexus" discovered by Nate Silver, but should also have physically mature arms. For college pitchers, this limit may come off, but high school pitchers could be as young as 20 years old and must retain this cap at any level, even the major leagues.

Triple A is like finishing school for pitchers. "Hitters know how to hit at Triple A," said the pitching coordinator. "That means pitchers have to know how to get hitters out at that level." This is still the focus, but in addition, triple-A pitchers should be allowed to extend their arms and find the level to which they can safely pitch the most. It is often said that we are "babying pitchers" in the modern era. Instead of testing them and hoping they can survive baseball's version of trial by fire, the progressive system will build them to their maximum effective use slowly, safely, and measurably. Pitcher A may only be able to go an effective 90 pitches while Pitcher B could go a safe 120. There will be the occasional pitchers who harken back to the high pitch counts of the '40s and '50s. When those are found who can safely go to 140 or 150 pitches per outing, we'll have proof rather than guesswork. For each and every pitcher, this system will allow them to give their maximum safe contribution, something neither the trial-by-fire system nor the strict pitch-count system does. The outliers will become known quantities.

The progressive system continues to work at the major-league level, both in an additive and diminutive fashion. Few pitchers come to the Majors as a finished product and few never make improvements. Some pitchers will extend upward from their call-up maximum and others, due to age or ineffectiveness, will lose some of their top-level pitches. The number also gets a bit softer, accepting the reality that a manager will sometimes have to go for a win or that a pitcher is "cruising." The system allows for an exception, but only once in a "set." Managers will now have to weigh one outing against not having that option in the pitcher's next four outings. Additionally, pitchers who exceed their number will have those taken off during the rest of the set. A pitcher who has a number of 120 could be asked to go to 130, but he would be only allowed 110 during the rest of the set. Bad outings, where a pitcher is pulled before his number, can bank up to five pitches, though this is discouraged. Basing that number on five-game sets, as well as the expertise of the pitching coach and medical staff, will allow fine adjustments in season and over the course of a pitcher's career.

We can look at two great pitchers of that era—Roger Clemens and Greg Maddux—as perfect specimens under the system. Clemens both extended further into games and continued to hold much of that pitch level into his 40s while Maddux has had to be more efficient underneath his pitch level (and appears to be losing some of it over the past few seasons). Certainly, there is a value in 80 pitches per game of a Greg Maddux–style craftsman. Our current system (or lack thereof) gives no quantitative tools; the progressive system shows in certain and easily understood ways how pitchers can safely and effectively be used.

There's an old tale about an ancient Greek wrestler, Milo of Croton, who used a primitive progressive system to gain strength. He put a newborn calf on his shoulders and walked around the field. Each day, the calf would get a bit bigger and the man would get a bit stronger. Eventually, he was carrying a cow and was as strong as a bull. (Part of the myth is that after four years, Milo slaughtered the cow and ate it all in a day, so there's some veracity concerns.) I wouldn't recommend the old calf over the shoulders trick, but who knows? Putting a cow on the back of pitchers might help to get the injury monkey off the back of baseball.

9

SCOUTS, UMPS, AND THE OLD SCHOOL

Well I did not think the girl could be so cruel
And I'm never going back to my old school.
— Steely Dan, "My Old School," 1973

You may not know the name Jon Poloni, but he was a legend among scouts. Today, after the wild success of *Moneyball* as both a book and a movie, most merely know him as "Fat Scout." In a famous scene, Brad Pitt as Billy Beane is trying to convince his scouts ahead of the draft that a catcher named Jeremy Brown would be a good pick. Poloni disagreed, and in his colorful vernacular, avoided calling Brown fat while still calling him fat. "He wears a large size of underwear," he is said to have said. The A's drafted Brown anyway and he of course went on to a Hall of Fame career with the team.

Wait, no he didn't.

While Brown made it up to the big leagues, a success by most measures in any draft, he was far from successful. On the other hand, Poloni had made his own case a few years before, pushing for a short, slim pitcher who was said to be too small and didn't throw hard enough.

Tim Hudson may not have joined some of his teammates in Cooperstown, but his career was just a bit better than Jeremy Brown's. When the A's won 20 games in a row, an episode that closes out the book, Hudson was a key part of that team while Brown was barely signed and headed to rookie ball.

Scouting may seem more art than science, but the tools, projections, and reports that scouts turn out often seem more like witchcraft. I can remember seeing a Dominican report on a 16 year old, described as 6'0, 125 and "malnourished." A scout watched the kid play, filed his report saying that he had "whip wrists" and could fill out. Five years later, Sammy Sosa made his professional debut and, a little chemical help aside, was anything but weak and malnourished when he fought Mark McGwire for the home-run title.

Omar Minaya was the scout who saw that skinny Dominican who everyone called "Mikey" and looked into the future, but when I spoke to him in 2004, Minaya had a harder time articulating exactly what it was he saw.

Minaya talked mostly about watching for signs of skills. While Sosa didn't hit for power, he had thick wrists and could move the bat well. He had good movement skills, especially laterally, which indicated he could likely play center field for a while, while the power developed. Minaya also remembered, later, seeing a picture of Sosa's deceased father, who was bigger, and while hardly well nourished himself as a poor Dominican sugar worker, had developed wide shoulders and thick legs.

Minaya was describing something that scouts innately look for and sounds like a baseball old wives' tale. Scouts will talk to younger players and ask how tall their father is. They'll look to see if the father became overweight in later life. They'll size up the mother and if she takes control of conversations. In essence, they're all describing genetics prior to the wide access to DNA sequencing, and without the thorny medical privacy issues that prevent wide usage of it even today.

Instead, scouts lean on a tool created in the 1930s by baseball legend Branch Rickey. While many things are credited to Rickey, he really did invent the scout scale, just as he created the farm system, and broke the color barrier by signing Jackie Robinson. (When in doubt, anything in baseball was invented by either Rickey or Paul Richards. More on him in a bit.)

The scout scale looks folksy, rating players on a scale from 20 to 80. Initially the scores were "on the zero," per Rickey. An average

MLB player was a 50. An everyday player or pitcher was a 60. A star is a 70. An 80, while almost unheard of, intimated a Hall of Famer and scouts were reluctant to ever use it in all but the most extreme situations. It steps down similarly on the negative side, though those designations are used less frequently and usually less descriptively.

In fact, this scale is pretty scientific, if also pretty basic. You may have already noticed that the three steps are essentially three standard deviations. It's reported that Rickey would go back and tally the scores and a scout needed to be within shouting distance of a true bell curve with his ratings or he'd have to explain to Rickey why he was, literally, behind the curve.

The modern scout hits the road in his or her area much in the same way that a scout did in Rickey or Richards's era. They go from game to game, practice to practice, with a knowledge base from years of experience in the game and a further base from scout school at the very least. Beyond that, they tend to know the right "baseball people" and often employ some "bird dogs"—locals that know the scene and the coaches.

The scout is almost always depicted as an older guy, often a former player or coach, with a Panama hat and a radar gun. It's a stereotype for a reason. More often than not, it's true. Scouts today have a radar gun and usually a video camera, but there's not a lot of modern technology that accompanies them.

Butch Baccala is one of those scouts who's been out on the road for years, working with several teams, including the Reds and Mariners. I asked him what tools he takes on the road. "Obviously anything you take to a game that isn't your eyes, your heart, or your ears—your instincts, your gut feel in my mind would be tools that measure certain skills," he responded. "I have my radar gun for pitchers' velocity, outfield or infielders' velocity, and a catcher's arm strength, so we get a lot there. The stopwatch measures speed even if it's not high-tech. These tools play a part in the data obtained over the years to come up with averages and what the real plusses are for these skills. I know in my head what those are, and some have changed, like what is a 50-tool fastball. It was 90 mph, now it's 92."

Baccala also notes that while he does the same overall evaluation on players, the position is also a key. "The importance of tools by position is more data driven now. Top three tools by position really guide things. It's almost a formula to come out with a player's overall potential, draft projection, and future major-league value, which is what we're all really looking for."

Scouts are also collection machines. Something like "signability"—how willing a player would be to sign with a given team—is key information and doesn't always come from the player or the advisor. Talking to coaches and teammates helps, but relationships are the key there.

Since the scouts on the road don't have much more than they did twenty years ago, the more high-tech stuff is done by the front-office staff, inside video rooms, and computer labs stacked with processing power. It's here where modern scouting is done, using video analysis and machine-learning technologies. They tend to use videos taken by the teams via multiple camera setups, like Synergy or PlaySight, that were originally intended for streaming productions. Instead, the video is able to be pulled apart and broken down, allowing the "number crunchers" (as scouts like to call them) to do more analysis and drill down.

By and large, the process is one of elimination. Bruce Seid, the scouting director for the Milwaukee Brewers until his tragic death, had a checklist that he'd look for with both pitchers and hitters. He told me in 2012 that he would tell his scouts to go out and see everyone, then start crossing people off the list. Did the pitcher throw hard enough? If not, cross. Did the hitter make enough contact? If not, cross. There was a list of about twenty things Seid looked for, from the "five tools" to confidence and grip strength, from "baseball IQ" to "did your brother play?"

Seid's list, broken down, looked at several skills—performance, projection, and mental skills—but also at genetics, confidence, and a response to stress. Yes, his list was folksy and full of hand-me-down wisdom, but there was a heart of science to it as well. What the modern scouting process has done is separate the two to some extent, allowing those with the ability to see into the future to separate from harder skills like video analysis, statistical projection, and psychological assessment.

Most of what Seid did translates directly from an old-school perspective to a new-school way. The language is different. The methods are different. The result is always the projection. If given new ways to create or fill in a projection, they have to be taken. That's especially true for scouts. Just as they adjusted to data when radar guns became ubiquitous, putting an objective if not always accurate piece of data down rather that their subjective 20 to 80 score for fastball didn't destroy the game nor the art of finding pitching talent. The same will be true for new tools like spin rate, biomechanics, or even genetic testing, if it meets its promise.

If an old scout feels undervalued, it's not because he's being replaced. It's often because he's refusing to adjust. New tools can help, not hurt, if only scouts can adjust.

The Umpires

In the modern game, some things don't feel like they are as valued. Scouts have felt that way—wrongly—but those that have adjusted to new tools have succeeded. The same might be true for another "old school" profession inside the game—umpires.

What's your earliest baseball memory? It might be your first major-league game, one you drove to with your dad or maybe the whole family went for the trip. More often than not, watching a game on TV—especially if something big happened—locks in. If you're one of those, what's that memory? What does it look like?

Is it in black and white, or color? (If yours is in HD, you're just making me feel old. Get off my lawn.)

What we see of the game is better now, and more. HD, 4K, and enhanced-reality broadcasts are giving viewers and even spectators more information than they've ever had. This is easiest to understand if we look at one of the simplest, most easily understood innovations: instant replay.

The history of instant replay is surprisingly opaque. It came about sometime in the 1950s, often cited as first used by a CBC hockey broadcast. The first slow-motion replay took a couple minutes, but allowed boxer Emile Griffith to analyze his knockout victory just after getting out of the ring.

Today, we take instant replay for granted, whether we're watching at home in multi-angle, super-slow glory or on the Jumbotron at the park. Still, it's a relatively recent innovation that has gotten steadily better as technology has improved in all phases.

Where we most notice instant replay is on an umpire's call. What we see in crystal-clear 4K, the umpire saw once at real speed. That strike that might be a bit outside due to the center-field camera's angle was judged by a 68-year-old man who answers to "Cowboy Joe." However, it wasn't until 2008 that MLB gave its umpires the ability to review a play.

The fact is that instant replay is the best tool available for the job. Those multiple angles and clarity allow a more nuanced view of a play, often resulting in a reversal. That isn't to say most umpires aren't good at their jobs, but the eyes of a man are no match for instant replay, let alone lasers, doppler radar, and machine learning. Simply put, instant replay is more accurate and more attuned that Cowboy Joe West's eyeballs. Period.

That's a genie that can't be put back in a bottle and with every passing day, it becomes more and more clear. The next step is a bit more extreme than checking the tape. It's robot umps.

Robot umps are likely not to be robots, sadly. Images of robots have gone from *The Jetsons'* "Rosie" to Arnold Schwarzenegger's T-800 Terminator, but what we're likely to get first in baseball won't even be as exciting or visible as a Cat Shark Roomba. (Serious question, who would win in a fight between a Terminator and a Cat Shark Roomba?)

Instead, we're more likely to get an evolution of the automated strike zone that has been tested at the minor-league levels. Much like K Zone, which is almost universally seen on television, the Robo Ump (as I'll call it) is a two-dimensional plane which determines a very simplistic outcome from the data: Did the ball break the imaginary plane of the strike zone as defined? In the current definition, it's a 2-D, not 3-D construct. Imagine a window pane set at the front of the plate, floating in space somehow. If the ball even tips the pane, let alone breaks it, it's a strike. If it misses, it's a ball. This sounds like I'm simplifying the system, but that's really it.

Obviously, there's more that goes into the multi-camera system that determines the position of the ball. There's a human element, setting the parameters for each batter. After that, it's essentially that window pane in the mind and a single, binary result—yes or no, strike or ball.

While there may be some Luddites out there that resist, especially the umpires themselves, this is coming quickly. The issue isn't going to be whether or not the system will be perfect—it won't—but whether it will be better. This is a question much like an autonomous car. Tesla built a car that can mostly drive itself, but when there's an accident, it hits all the papers. When a robot ump misses a pitch—or appears to because of an offset camera—the Twitter wags will come out in force, the old-school *I told ya so's* will be just behind them, and the umpires who will be on the field as well will just throw their arms up, reminding people that this robot took my job, can you believe it?

However, this won't be an egregious error. It might miss up, or miss down, or even have an operator error. It won't be an Eric Gregg pitch in the opposite box. It won't be an Angel Hernandez ball that's right down the middle. It won't be an inconsistent zone that shifts from inning to inning, or gets pulled out wide by Tom Glavine nibbling the edge again and again, drawing the ump's eye out. It won't get tired or worry about missing a flight if this game goes extras.

Instead, it will be small errors and likely well under the error level we see. Again, umpires are very good at what they do and we regularly see accuracy of balls and strikes at about 93 percent. This is despite the fact that each umpire is setting their strike zone by experience and in their mind rather than by a true, objective measure.

Baseball as a sport moves at a glacial pace. Change is often made slowly and only after it's been discussed ad infinitum and then just a little bit more. There's almost no chance that MLB will drop in robot umps until they're sure that they're not just a little better than human umps; they'll be very, very sure that it's an order of magnitude better. MLB is already testing this at lower levels and has been iterating tests using both its original Statcast system and the updated Hawk-Eye–based system that came in during the 2019 season.

The effects of this kind of system are almost impossible to track. Baseball experimented with the system over several levels, including the Low-A Southeast minor league. Since they changed levels and teams, it's nearly impossible to take this league and compare it to the Florida State League, which was previously High-A, or the South Atlantic League, which was previously Low-A. In 2019, the Low-A South Atlantic League had about a five-point batting average difference to the 2021 Low-A Southeast (.234 versus .239). In terms of strikeouts and walks in those years where the system was used, the raw numbers aren't comparable and even the percentages are very different—higher walk percentages—in ways that don't seem to match up.

The New Scouts

It's a hot Friday in July and I'm sitting on my back patio. My laptop is in my lap and a cold beer is beside me, sweating almost as much as the players I'm watching play a baseball game. It's been possible to watch MLB games online for years, with the innovative MLB TV app light-years ahead of other leagues. Baseball was so far ahead, they sold the subsidiary that did all that to Disney for billions, a big profit for the team owners.

The game I'm watching in one window while I type this up is between elite 16 and under players. One is a team of all-stars from Florida while the other is from Illinois. Prep Baseball Report, a scouting service that ranks players and organizes tournaments and showcases throughout the baseball season, is putting on its Future Games and I'm watching through a site called PlaySight. The multi-camera setup isn't what I'd expect from MLB or ESPN, but it's very watchable.

In another window of my Mac, I'm chatting with two scouts from different MLB teams. One is on-site in suburban Atlanta. When PlaySight shows the third-base side camera, I can see him at the edge of the frame. The other isn't there, watching through PlaySight like me. Both are scouting the same two kids at the same time in very different ways.

There are positives and negatives to both ways. The key difference is that there is a certain feel and openness to being at a game. For the scout that's at a game, he can see how a player reacts to things that might not be caught on camera. Is he sulking about a strikeout or supporting his teammate? Is he talkative or focused? "Fielding is the key for me. I can scout a pitcher or hitter with video, but I need to watch the fielder live or at least have full field shots," said the scout who was at the game. "I want to see how he adjusts pre-pitch. Is he moving? Does he get caught out?" The scout that watched on video acknowledged that, but says he can fill in the gaps. "If he's making the plays he's supposed to, I'll know he's fine pre-pitch. The result of the play will tell me things about positioning and I can see the things I care about in fielders like arm strength on plays."

Watching a game on video is not like watching a telecast, though there are similarities. "It's gotten so much better in the last two years," said the scout watching from his home. "The cameras are better and there's more of them. At most tournaments, I get a pretty standard view—home and center—but I'm seeing more with a side view or a wide view for defensive plays. With a lot, I can find highlights or different angles on YouTube but it's usually someone with their phone. Then again, I had one kid I scouted in 2020, when we couldn't travel and his mom was taking 4K video and editing it that night. The kid wasn't that good, but I wish I could have drafted his mom."

Speaking of moms, the scouts also disagreed on home visits. While the at-game scout thought he could get more with body language and "looking them in the eye," the at-home scout thought there's probably some slight advantage to in-person, but "we're all living on Zoom these days. I think I can get more from regular contact than the one time I drive through north Louisiana."

Both agree that the explosion of data and video isn't going to slow down. The at-game scout said that he'd have to wait a day to get TrackMan data and that teams didn't always share that. "More top facilities do. Where I'm at now has them on four fields and those cost, what, 20 grand each?" Rapsodo data isn't given from in-game, one of the limitations of the system, but is regularly used in bullpens and even scrimmages at many colleges.

In the end, both likely have a place going forward. With video, there's a chance to see more players, saving the in-person visits for the players that are rising to the top. As more and more data enters into the scouting discussion, there will end up with a multi-tiered approach and hopefully, the old and new coming together to find the new talent baseball needs.

Conclusion

There is nothing more tired than the whole Scouts versus Stats, Old versus New argument. I know, I helped perpetuate it, putting Michael Lewis, the author of *Moneyball*, on a radio show opposite Tracy Ringolsby, the Hall of Fame baseball writer that positioned himself as the champion of the old school at the time where *Moneyball* was new and controversial. It was less about a book and more about a metaphor.

I thought Dayn Perry, a great baseball writer himself, had put the whole thing to bed in an essay back in August 2003 where he came up with this great line: "A question that's sometimes posed goes something like this: 'Should you run an organization with scouts or statistics?' My answer is the same it would be if someone asked me: 'Beer or tacos?' Both, you fool."

Sadly, there is no shortage of fools.

Baseball doesn't need fewer scouts or fewer umps, but if technology makes it possible for fewer scouts to do the same work, then yes, the business side of baseball will take care of that. If umps are given tools to do their jobs better, to be more accurate, and to move the pace of the game along, why would there be any argument?

What baseball needs is great scouts and great umpires that use the best tools to do the job they've been asked to all along. If science can help them do that, I think the old school will do just fine in the new, scientific world of baseball.

Now, I want a beer and a taco.

10

THE MIND AND BODY

Is the body a temple, or a machine? Given how most baseball players right up to the modern day have treated their bodies, neither is really the answer. Mickey Mantle said, "If I'd known I would live this long, I would have taken better care of myself," and a lot of players could say exactly the same. Baseball is a game of failures, rainouts, and long bus rides, which make for a lot of depression, drinking, and lack of sleep.

Baseball's always been a balance between the angel and the devil, with talent as the deciding factor. Mantle could drink until dawn, roll into the Bronx, and hit. Not many can do that, largely because not many are Mickey Mantle.

For modern baseball, it's one thing to train the body to pitch, to hit, to run, and to field, but there's even more to it. One key that we've seen is professionals bringing non-traditional techniques like meditation, yoga, mobility training, and more into a modern scientific context. It's easy to see the science of using something like a Theragun, which is now nearly ubiquitous, a Marc Pro electrical stimulation machine, or something like a pair of Normatecs, to help the legs recover with compression and massage. It's less clear when you see an athlete or even a whole team lying on their yoga mats in savasana, the "corpse pose" that looks like everyone's just taking a little nap.

For the practitioners around baseball, fusing the conservative world of baseball with the more progressive worlds of ancient practices and modern variants, it's all about the individual. Bridging that gap between two disparate worlds is Steph Armijo. A passionate baseball fan who is also one of the most sought after yogis, Armijo started her company, Yoga42, after spending time in

MLB's Stats department. She's one of the first yogis to use wearable sensors to help with yoga practice, which is going to explode in use over the next few years.

People have a hard time imagining baseball players doing yoga, and that innate resistance to something outside the norm comes out regularly. "Did Nolan Ryan do this?" is a question many will get, but Ryan was always willing to work outside the norms. It might not have been yoga, but doing what was best for his body was always within the realm for Ryan. It's easier today, but not by much. That makes the job even harder for Armijo, breaking through that innate resistance to change, let alone some "California hippie stuff" like yoga.

Armijo has seen this firsthand. "Front offices and coaches tend to do what they have always done. I had an MLB manager refuse to join in during a meditation for the entire team. He told me, 'if I wanted to take a nap, I would have stayed home on my [bleeping] couch.' When there is not support or buy-in from the top down, it is not a priority."

What has made Armijo and a few others like her successful has been blending the old with the new and giving it a specific baseball flavor. Even something as simple as her company's name shows she's got baseball close to her heart, and woe be to those who question her baseball bona fides as a woman. I know few more passionate and knowledgeable fans.

That baseball flavor is the secret sauce. I asked Armijo how she's adjusted what she does to who she works with. "Traditional yoga is taught in a group setting, with the same poses and sequences prescribed for all students. Yoga42 provides individually tailored yoga classes specifically for athletes," she explained. "I customize my yoga classes for each athlete based on their mobility needs in the short and long term. Powered by computer vision algorithms from standard video captured during a quick mobility scan, I compare an athlete's pre- and post-yoga range of motion to their long-term baseline. This gives me a tangible way to show athletes where they can improve mobility and asymmetry. Furthermore, I've been able to amass a biomechanics library of yoga postures that helps me map yoga poses to an athlete's biomechanical needs. Using that information, I am able to sequence my yoga sessions for each player in a way to keep them safe and optimize movement patterns."

With this specific and scientific approach, Armijo has been able to break through the resistance. It becomes a "try it, you'll like it" thing for most. "At this point, most baseball players have been exposed to yoga in high school or college and they know it can help them feel better. Yoga and meditation also have strong endorsements from world-class athletes across all sports, and the mental aspect of baseball is now widely accepted," she explained. "Yet teams still aren't prioritizing these practices in a routine manner. A lot of MLB teams incorporate aspects of it, but in a limited manner that is often fused into their strength and conditioning routines. Given the limited exposure, players do not receive the recovery and mobility benefits they would experience from a more regimented yoga practice. To see benefits, dedicated yoga and meditation should be practiced and offered multiple times per week during the season. With 162 games plus spring training and postseason, there is a lot of wear and tear on the body. Using yoga and meditation as a recovery practice has physical and mental benefits for baseball players."

With Armijo's baseball-specific practices, the athletes can see immediate effects in their mind/body connection. "Being present during the most important moments keeps athletes alert, healthy, and performing at their best," she said. "Elite athletes can easily sense when something feels 'off' and often make appropriate adjustments quickly. I worked with an MLB outfielder that began to notice every sore muscle, tweak, and tightness. He was able to address these fatigue markers with his medical staff on a daily basis.

When you start moving your body in a complementary way to your everyday activities, you can tap in and notice changes quickly. If you are not aware of how you move and how your body feels, you are more likely to overwork or stress your body and that could come at a large cost for the athlete and team."

That awareness is difficult for athletes, but with demands on their time as well as their bodies and minds, it can be the difference between sustained success or failure for some. All athletes will slump, but those that can adjust to reduce those slumps are bound to have more success. That alone can surprise players. "Players are surprised they can slow down and connect the 'feel' element. Most players immediately say they feel better and move better," Armijo told me. "It's difficult to get some athletes to down regulate and spend time breathing, but the transformation is almost immediate. I worked with an MLB pitcher that wanted to sweat, every workout. He was convinced he needed hot yoga to make the practice worth his time. Yoga and meditation are experiential—players typically need to try it before they will buy in."

Beyond the mind/body connection and presence, there are physical effects as well. "Moving with breath is not a completely intuitive action. We all hold our breath, and to not hold your breath in challenging scenarios takes practice," she said. "The more an athlete can stay present and breathe while moving, the less likely they are to get injured. When we hold our breath, our nervous system goes on high alert to protect the body. So if you make a sudden move, like a dive or awkward reach for the ball, your brain recognizes [it] as dangerous. If you are not breathing, you are more likely to injure yourself."

Armijo continued. "Another major benefit is better, more restful sleep. The baseball schedule is intense, and night games plus constant travel are not ideal for the body to heal and recover from fatigue. Meditation and breathing exercises can be tremendously helpful for players to fall asleep quicker and feel more rested when they wake up."

It is still very early days for practices like Armijo's yoga but as a leader in the field, she sees the benefits expanding. "Breathing techniques and meditation can be done in nearly any complementary scenario," Armijo said. "The more it is practiced, the easier it is to tap into that mindset. One thing the industry is seeing more of, are dedicated live-in personal trainers and massage therapists that even travel with the players to augment their needs outside of the team-offered services. Players are starting to do this with yoga teachers and I expect this to become a more frequent service."

Armijo's yoga is easy for most to envision. We've all seen or even tried yoga, and while it's more than poses and flows, it is more familiar to most than "mobility training." The fact that it exists or that professional athletes will reach out to a "performance recovery coach" bears some explanation, so I asked Sarah Howard, one of the top performance recovery coaches working in baseball, to explain exactly what she does.

"I help athletes stay healthy and recover better," explained Howard over FaceTime from her base near Los Angeles. "Athletes have had years of practice at working hard, grinding, and going at full throttle, but they haven't been taught how to rest, recover, and take care of their body. They need to know how to do this and they need to know how to do it on their own. I am working on changing this entire scenario, by teaching athletes how to heal their own body and be able to do that in a budget-friendly way, anytime, anywhere."

Howard came to her work in much the same way as Armijo. "I have always been an athlete and played multiple sports growing up, then tinkered with things like rock climbing, road biking, and olympic lifting after college. My body tends to be stiffer and my mind has always been active," she told me, sipping her drink from a bottle. "I realized how much yoga, and then the Roll Model Method, helped my body and mind feel better and allowed me to stay healthier. It dawned on me that I was never taught, through all my time playing sports from a young age into college, how to take care of my body and my mind. I was taught how to train, but never what to do after. I saw the same issue with athletes now, watching the injuries pile up season after season. So, I decided it's time to change the game, and Elite Mobility Training was formed!

Howard made a smart move, linking up with one of the top pitching gurus early. "When I moved to Los Angeles and started my company, I didn't know anyone. I began reaching out to people in the sports industry and I connected with Alan Jaeger," she said, referring to the highly sought-after pitching coach. "We quickly became good friends and as they say, the rest is history! I got linked up with Jack Flaherty (now of the St. Louis Cardinals) and began to put more of my work on social media. Through that and word of mouth, I became known in the baseball world. The combo of therapy ball work, both large and small, and yoga is ideal for reducing the common injuries that plague baseball players."

Think of mobility training as something descended from yoga or Pilates, but more specific to sport. Howard has a focus on the core, which has become a buzzword since Mark Verstegen launched it into the world about twenty years ago. The core is simply the center of the body, largely the abdominals, obliques, and supporting muscles. I once heard a pitching coach describe most youth pitchers as "two steel plates with Jell-O in between." Young pitchers develop their cores late, after their hips and shoulders, which are largely bony in nature. As we discussed in previous chapters, that hip/shoulder separation and the ability to control the body is key across most activities of baseball, so focusing on the core as Howard does can make huge differences.

I asked Howard why core work was so important, especially for baseball players. "Good movement starts from the middle—the abdominal area—which is why tools like the Coregeous ball [a small inflatable ball used to roll the body against] are such a huge part of my training. Why? The branches of a tree are only as healthy as the trunk of the tree. In terms of optimal movement, we are no different. A strong core is great, but stiffness here impedes shoulder, thoracic spine, and hip mobility, because it's all connected! The different balls with their sizes and hardness helps to de-stiffen and hydrate the muscles of the abdomen, freeing up movement, breathing capabilities, and reducing pain."

While those not familiar with this kind of training might think rolling around on balls looks anything but scientific or therapeutic, they'd be very wrong. Howard detailed just some of the ways that mobility training works and the science behind it. "Self massage with therapy balls and the Coregeous ball stimulates various mechanoreceptors. When stimulated, these sensory nerve endings can release tension, lengthen myofascia, and reduce muscle bracing. As we say in this world, 'motion is lotion' and rolling with these balls also helps water molecules bond with collagen fibers which hydrate tissue and fascia."

There are plenty of studies that back this up as well. "Two studies done by Dr. Robyn Capobianco showed that rolling the gastrocnemius (calf muscle) on a soft rubber ball was able to offset a stretch-induced force deficit, while also increasing torque, range of motion, and force production. These results were demonstrated by both college and middle-aged adults. The results for middle-aged adults were extremely interesting, as their ankle dorsiflexion—the lack of which often drives slips and falls—increased by almost 25 percent."

Any athlete is going to be looking for immediate benefits and Howard says she can see players getting results in their first session, but it goes beyond that. "The immediate benefit is better movement and reduced pain. I can have athletes working on the balls for five minutes and they gain movement, whether it's shoulder mobility, thoracic spine rotation, or hip mobility. Athletes can move better, feel better, and are able to keep doing what they love."

Where there's even more benefit is the long-term practice. "I've been doing this work for about six years, so I don't have an exact answer for long-term benefits," Howard explained. "What I can tell you is that every former player that I have talked to or worked with tells me two things. First, they wish they started this work when they were playing and secondly, they tell me how much discomfort and pain they are in from their playing days. My hope is that by athletes starting this work now, their body will be less trashed once they retire so they can live, play, and thrive in their post-playing days."

Clearly, this kind of training is important and not overlooked, but individualized and very specific programming is why coaches like Howard are in such demand. Despite this, the costs are reasonable. "Theraguns and private sessions with a massage therapist are amazing, but they are expensive and sometimes, more often that not, aren't in the budget," Howard explains. "Elite Mobility Training offers online training programs that range from $39 to $89 and the cost of balls is less than $40. For less than $100 you can take care of your body from head to toe, every day."

She's also seeing more teams get involved as the practice is more accepted. "I think we are going to see a lot more of this kind of work implemented into team time. Athletes are becoming more vocal about stress and mental health," she said. "Living in a hyper-aroused state 24/7, which is how athletes live now, isn't good for the body or mind. Using practice time to turn on the off switch will become more and more popular over the next five to ten years. Perhaps even having a person on staff who does my kind of work will be huge, for both the players' and the coaches' health."

Supercharging Sleep

Perhaps the final frontier in this kind of mind/body connection work is emphasizing the importance of rest, recovery, and sleep for athletes. Over the past eighteen months, I've worked closely on an ancient technique called yoga nidra, which is often referred to as sleep yoga. While I'm hardly an expert in the way that Armijo or Howard are in their areas, the use of yoga nidra in sport is so nascent that finding an expert on their level is impossible.

I'm reminded of my 2004 book, *Saving the Pitcher*, in which I discuss using this strange implement called a kettlebell. At the time, it was largely referred to as a Russian kettlebell and popularized, as it is to this day, by instructor Pavel Tsatsouline. Kettlebells are now widely used in baseball, as well as many other sports. You're likely to find them at a retailer like Walmart, alongside the yoga mats!

Yoga nidra (or just nidra, as I'll refer to it from this point) can be done without any special equipment, though a yoga mat can be useful. The entire practice is done lying down and there's little if any movement. The practice isn't easy and is intensely mental. The focus is not falling asleep, but getting to a state of conscious awareness at rest. Have you ever had one of those days where you didn't do much physically, but there was a lot of mental work or stress and you found yourself oddly rested physically, but frazzled and burned out mentally? Nidra is the opposite of that.

Athletes especially can be very physical, for obvious reasons, and just lying still can be intense. I've watched athletes I've worked with try hard not to fidget and after just moments, they pop up. Their sense of time and even place can be altered because of the intense effort.

On the other hand, some do relax, forgetting that conscious part. The goal of nidra is not to fall asleep, though many will do so and when it happens, it's a signal that the body needs more sleep. If the athlete ends up with a nice thirty-minute nap, there are many worse things they could be doing.

Finding the right balance, being able to calm the body and allow the rest part of "restorative" to start, even beyond the slow pace of restorative yoga, is key and the most difficult part. After a practitioner can find that state regularly, it's amazing how quickly they can again find that state when it's needed. Pitching coaches will often tell athletes to relax or "take a deep breath, buddy!" but have they ever practiced that? Some athletes benefit from biofeedback, feeling their heart rate and respiration drop as they sink into the consciously aware state of nidra.

Studies have shown that the practice of nidra has benefits beyond simple rest or even meditation. Finding that deep set of consciousness can trigger reactions in the parasympathetic nervous system, which controls everything from digestive health to sexual function. Changes in heart rate have been observed, plus some biochemical changes, including increased production of dopamine and melatonin, often in short supply due to our connected world of screens and blue light.

Most modern forms of nidra have been developed by Dr. Richard Miller and have been approved by the military as a Tier 1 tool for pain management. In my practice, athletes who are generally overstimulated and dealing with lingering soreness or even pain can benefit from a nidra practice with very little active guidance.

However, there are more advanced practices once the simple conscious awareness is reached and can be held. The first is called "sankalpa," which is often translated from the Sanskrit as "heart's intention." It can be as simple as setting a positive goal. Most athletes are exceptionally goal-oriented, but have seldom worked to make that both positive and internal, with sports having so many external goals, like wins, more velocity, or lifting heavier weights. A pitcher's sankalpa isn't about more velocity, but can be internalized.

A pitcher hoping for more success can see his sankalpa as "I am a winner" or "My body will help me win."

Moving forward, I find one of the most exciting areas of nidra to be the rotation of consciousness, or annamaya kosha. This refers to understanding the direct relation between the body and mind by moving the focus of the conscious awareness through the body. Rotations can move through several patterns, from spot to spot without actual physical movement. It is only the consciousness that moves.

Let's try a simple experiment, assuming that you're not in the proper state. Close your eyes and think about your index finger. Focus all of your attention on that finger, but don't move it. Now move to the middle finger. The ring finger. The pinky finger. Did you move any of them as you thought about it? A rotation is much more complex, but also as simple as the conscious focus moving and learning to feel how your body is connected, both physically and mentally, and in a sphere best described as a flow of energy.

Nidra is simple, but not easy. As I said, all equipment is optional and you could simply lay in bed after downloading any of a number of podcasts or YouTube videos. I recommend the work of Supernova Yoga Nidra by Shannon McPhee, but your journey is your own.

The world of the mind/body connection, of yoga, mobility, and yoga nidra, as well as other connected practices such as body work, Pilates training, or others, may seem entirely outside the world of baseball and at times, the world of science. Instead, we're finding that the connections between the mind and the body can often not only help performance, but keep players from breaking down in the first place.

While these practices are not yet widespread in baseball, especially at lower levels, trust me—you do not want to have a closed mind when it comes to a new way to improve.

11

INJURIES

If there's any way in which baseball could get better, and wherein a team could find one of those asymmetric "Moneyball" advantages, it's in injury management and prevention. In 2019, MLB teams lost $850 million on players and 47,000 days, which is almost 1,600 per team, though the injuries are not equally distributed. (That data is from Spotrac, and sources inside baseball have corroborated its accuracy.) The previous decade saw similar yearly losses approaching a billion dollars, and if you don't either gasp or hear Dr. Evil's voice in your head when you hear those numbers, you're reading the wrong chapter.

The Doctors

Birmingham, Alabama, isn't where you'd expect the mecca of sports injuries to be located. Downtown Los Angeles is much more Central Casting for this, but those two are the centers of the sports world when there's an injury.

In Birmingham, Dr. James Andrews and his slow southern drawl and folksy manner belie the fact that he's been a top surgeon for everyone from Roger Clemens to Adrian Peterson. Originally from Louisiana, Andrews might not look like an SEC pole vaulter, but he was. He parlayed that athletic success into academic success, ending up in medical school, then becoming the right hand of Dr. Jack Hughston, the top sports surgeon of his day.

In the early 1980s, Andrews saw a chance to change sports medicine with the use of the then-new arthroscope. Hughston was a bit more conservative and Andrews ended up heading northwest, settling into an old hospital that was retrofitted for his use on the south side of Birmingham, near the University of Alabama-Birmingham campus, and within walking distance of Legion Field, then the annual home of the Alabama–Auburn football game.

One only needs to witness the passion of this in-state rivalry to fully comprehend it, and yet Dr. Andrews is the team physician for both teams. It's like Nick Saban coaching both sides—unthinkable, but Andrews and his associates are that important, covering not only Alabama and Auburn, but several other area colleges, the Washington Football Team, the Tampa Bay Rays, a clinic near Pensacola, Florida, and a new one north of Dallas.

Across the country, Neal ElAttrache lives in the shadow of the Hollywood sign and looks like a movie star. The Pittsburgh native ended up in Los Angeles to work under another of sports medicine's greats, Dr. Frank Jobe. Jobe created Tommy John surgery and with his partner, Robert Kerlan, created the Kerlan-Jobe Clinic. Kerlan had been the first team doctor for the Los Angeles Dodgers from when they initially moved west, handing off the team to his partner when his arthritis, which often necessitated the use of crutches, became too much for him to continue.

ElAttrache has worked on nearly as many athletes as Andrews, with his client list including Tom Brady (knee), Kobe Bryant (Achilles), and hundreds more. His research and work on technique have led to better outcomes for Achilles repair, labrum repair, and Tommy John surgery, where his variant on it has become the most widely used. He took over the Dodgers from Dr. Jobe, and then the Los Angeles Rams.

It's not just that Andrews and ElAttrache are at the top of their professions, but they largely help mold the rest of the industry, not just in the NFL but in sports medicine as a whole. Andrews helps lead the Injuries in Baseball Course, an ongoing series of annual conferences that are going on thirty years and draw in the top names in sports medicine, from doctors and athletic trainers to researchers and even the occasional writer. ElAttrache leads his own conference as well, originally established by the late Dr. Lew Yocum, another Jobe protege and a longtime Angels doctor.

The same is true with fellowships. Getting in for speciality training at Andrews or Kerlan-Jobe is a career maker. Around the league, top doctors like Dr. Tim Kremchek (Cincinnati) and Dr. Keith Meister (Rangers) came out of Andrews's program. Dr. Jeff Dugas is at the leading edge of elbow surgery, one of the top advocates of a new technique that could replace Tommy John surgery for many. ElAttrache hasn't been at the helm as long as the others, but he remains a strong branch of the Jobe tree.

The fact that both Andrews and ElAttrache are household names for, if a bit feared by, average baseball fans tells you just how big this problem is, and one look at their waiting rooms, filled with players from all levels, suggests that no one's implemented a solution. Yet.

Getting to Zero

Unicorns. Hobbits. An injury-free season in baseball.

These were all considered mythical creatures in the game until 2019, when a professional team in Taiwan, the Chinatrust Brothers, had a full season with no soft tissue injuries. Yes, it's a bit of a technicality to call it an injury-free season, but no amount of planning, hard work, or technology is going to stop an injury like a broken bone after a hit-by-pitch or something fluky, like Clint Barmes falling down stairs after trying to carry a deer he'd killed up to his apartment. (True story.)

What Gary McCoy, an Australian sports scientist who has worked with several major-league teams and helped incubate some top technologies like Catapult, Whoop, and Kinetyx into the market, was able to do with the Brothers is nothing short of astounding, but it wasn't luck.

Today, Gary lives near Phoenix and is working on some amazing sports technologies, but that feat of "zero injuries" is inspirational and aspirational. What always strikes me when talking to him is that the first thing Gary talks about are the players he works with. Yes, that's unusual.

"Just like an architect doesn't take credit for the structure and the internal design that makes a house a home, my job was to provide the blueprint design and then audit the team in place to be able to deliver on that design," he said in his still-thick Melbourne accent. "I educated the team. I was always available as a resource. I supported every effort our staff made. I gave room for creativity if aligned with the central vision. With a larger staff, this was a lot of early work."

It's a bigger team than many would expect. "In the organization under my direct supervision were the following cast of characters," he said, counting them off on his fingers. "Head athletic trainer and six more athletic trainers across the organization. Three strength and conditioning coaches. One soft tissue readiness practitioner—note I didn't say massage! One team orthopedist, one team psychologist, one data scientist, and an organizational nutritionist. All these areas had to share transparently and align like an F1 pit crew."

McCoy continued, detailing how he organized that team. "We had weekly meetings on roster evaluations and then came to educated agreements on directives. Over four years, I said less and less in these meetings and the key in-season role for me was coach communication. I had such good rapport with the coaching staff, so I could explain adaptive load and constructs of recovery and readiness ahead of time. That made it so I could walk up to the manager in game and say, 'Hey, this pitcher is running in the red' and he knew what I meant, knew what the consequences were. Not only that, I trusted him to do the right thing rather than relying on us to put the player back together when he was overused."

McCoy did see changes in that team and how things morphed as the system became more ingrained. McCoy pointed to one particular need that came up. "Our biggest change after year one and heading into two was the identification of the need to hire a strength coach with massive attention to individual movement detail and an understanding of vector-based training, especially the properties of kinetic decelerative principles," he said. Yes, that's as complex and specialized as it sounds. Instead of hiring someone, the team provided the resources to let the existing strength and conditioning coach educate himself to fill the need. "[That coach] basically relearned his craft and was the transcendent layer that achieved the next massive lowering of soft tissue injury," McCoy raved.

As for those vectors and decelerations, McCoy's system has a focus on balance. One simple thing he does is have batters take swings from the opposite side after batting practice, to gain some feel for the imbalance. If you've ever tried to do something with your non-dominant hand, you likely have the same issues players do. Though the non-dominant hand can be trained, the more likely course is just to do everything one-sided. Using a pen or scissors on your dominant (likely right-handed) side isn't going to create many unidirectional compensations the way taking a lot of baseball or golf swings will. The same holds true for pitchers, though McCoy uses some specific exercises rather than trying to turn everyone into ambidextrous relievers.

Asked about how to succeed at reducing injuries, his answer focuses on the system again. "It takes an organization. They have to give a dedicated and skilled professional two things—first, the responsibility to own the numbers, to own player injury, and to own player performance. Second and directly attached to this is the authority to make the decisions required to put the right team in place with the right skills and the right personal accountability. Hire and fire staff justifiably. The evolution of the athletes and the evolution of the sport depends on this."

As easy as that sounds, McCoy sighed when I asked him why this isn't done. "Most dreams die in the mind of the dreamer long before they make it out into words, lists, or actions. We are often our

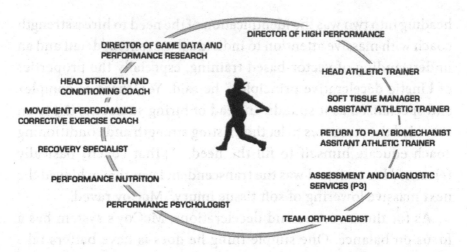

McCoy's performance flow chart from the Brothers. (Courtesy Gary McCoy)

own central governing system. For anyone to take a step back and realize that a different process avoids the current injury insanity takes courage. I haven't seen that courage anywhere yet. All I see is the same systems achieving worse results. Fans and owners pay that price, and at the center of paying that price is our MLB athlete. Keeping them on the field changes the game."

I wanted to drill down on McCoy's definition of preventable. Which are the injuries he saw as the most preventable? "Easy one," he said quickly. "Hamstring and oblique strains. I haven't had one of these in twelve years working with teams directly. For some reason in baseball, we chase pain. We treat symptoms or we treat the artifact of a misaligned kinematic stress. The answers are right under our nose 99 percent of the time." You might note here that hamstring and upper leg injuries were the single biggest injury in MLB in 2021, up almost 40 percent over the last complete season in 2019.

It was time to hit him with the key question—could a major-league team do this?

"Zero overall injury is fantasy land for all of us, but zero preventable injury can be achieved," he told me. "We have to separate injury surveillance into two channels—preventable and non-preventable, which isn't hard to do. The preventable are all the soft tissue, chronic injuries that occur in baseball. We need to run as much data against each injury, by athlete, to understand the pattern that potentially is at cause for this."

McCoy's research has been focused on this. "There have been four primary reasons for preventable injury across multiple sports. In order, these are overtraining, under-training, lack of specificity, and lack of variation." Those are really two "flip side" areas, so McCoy is advocating a middle path, one that is individualized. Creating a long-term plan for an athlete or putting them on a single, unadjustable training program is going to fail, he says.

"I'm perplexed when asked to write a baseball conditioning guide," he laughed. "I'd only ever be 60 percent accurate due to the innate variability of human physical systems. The other 40 percent is all to do with the individual! We need twenty-six individual programs for a big-league team."

I had to put him on the spot and McCoy didn't flinch when I asked him what kind of results he thinks a team could get in year one. "Put the right system in place and it's huge. If a team simply did this much in year one—October to October, because you have to include the offseason—they could see a 50 to 65 percent change in preventable injury. I saw 75 percent in year one with this application."

So what's stopping them? McCoy had a ready answer, though he tried to avoid too many generalities. "There are some common themes that run through MLB organizations. I have three that repeat in cycles for me in discussions with many clubs," he explained. "The first is knowledge. Owners and GMs don't know how to measure the performance of medical staff effectiveness.

This should be simple. First, reduce man games lost year over year, at least until you get to a regular position well below the league average. Second, benchmark your team versus all others and answer whether you're getting the players you lost back on the field quicker and without re-injury. Finally, set and measure what your key performance indicators are and focus on the ones that are physically governed." Those would be things we've talked about in previous chapters like velocity, spin rate, ball exit velocity and launch angle, pop times for catchers. Perhaps you suddenly see how important all this is, how holistic. Measure those, as you should be for other purposes, and use it to help guide the sports medicine and sports science part of a team.

To explain a term McCoy used, "man games" or "man games lost" is the common term for how injury losses are measured. A man game is one player losing one game. It's possible that two different players miss the same game, which would be two man games lost.

McCoy also thinks there are a couple other commonalities, and they go together, unfortunately. "Fear and history. Baseball is history. I can't tell you how many conversations in baseball start with 'back in my day . . . .' When a coach starts that way, he's resisting change. It's great that the history of the game is so much a part of the American Dream, but that same history shouldn't be an anchor on progress."

He continued, "Fear is just human. It's fear of something different, fear of something new, fear that someone will be replaced, fear that someone won't be able to stay in baseball, or all the other things our brains use to hold us back. There's a baseball saying—you can't steal second while you have one foot on first—and that holds so true for how baseball resists change and progress."

But billions of dollars have been lost and will continue to be until we see that change. McCoy told me that bioanalytics is what he thinks will become the Moneyball of the 2020s, an advantage that can be exploited by a team bold enough to do things differently. Does baseball have that team now or an owner willing to push it?

Athletic Trainers

Steve Cohen owns the Mets and runs a hedge fund called Point72. The site WhaleWisdom, which tracks well-known investors like Cohen, says this about his fund: "Point72 Asset Management is a hedge fund with 17 clients and discretionary assets under management of $117,537,549,106." Seventeen clients. $117 billion on hand. I'll just let you come to terms with those numbers, while I tell you their biggest holdings are in companies like Uber, Western Digital, Baidu, and Visa.

Imagine Cohen has an investment in one of those big companies and they come to market and say, "We lost a billion dollars this year. Some of our top assets failed and our executives just couldn't fix it. It's not so bad, because we did it last year and we'll probably do it this year."

My guess is Cohen's response would be something along the lines of Bobby Axelrod, the anti-hero at the center of Showtime's *Billions*, who is said to be in part based on Cohen himself. I don't think it would be to shrug his shoulders, take the loss, and watch his hedge fund fall in the rankings, rather than trying something that is proven to reduce injuries by as much as 50 to 75 percent, in the specific sport!

That's basically what all owners—not just Cohen—do when they watch the kind of losses happen to players under contract. The days lost for players amount to billions over the course of a few seasons and yet nothing has changed. Few teams have made a huge investment or done much to improve, expand, or be innovative when it comes to medical statts.

Just look at the research. Andrews and ElAttrache have done hundreds of studies between them, many focused on baseball injuries, but MLB itself barely budgets a million dollars a year for research grants in this area. No team has a budget that reaches into seven figures. The difference between sports medicine, sports science, sports performance, and associated staff between MLB teams and an average top-tier European soccer team is easy. It's not close.

Quick—look up your favorite team and tell me who their lead sports scientist is. Surprised? My guess is that twenty-eight out of thirty of you had to look at the whole page, wondering if you'd missed it or if there was maybe an alternate title. And don't think it's just not listed.

I couldn't discuss injuries in baseball without acknowledging the real heroes of the game—athletic trainers. As with doctors, it's difficult to find out when the first AT came into the game, as the profession evolved, gained allied health status, standing alongside professions like physical therapists, and becoming more and more technical.

Athletic trainers have to pass testing boards, put in thousands of hours of work, and be able to work with doctors on all kinds of medical issues. When you see a player go down on the field and a couple guys rush out to help him, that's the athletic trainer, yet most fans can't name the guy who's in charge of keeping the team's players healthy!

While you will find two, maybe three ATs listed for each team, plus one at each level of the minors, think of the man hours that go into doing all the preventative work, prepping players, doing pregame routines that can include therapeutic modalities or massages, or even just taping ankles. Then they're at the game for hours, followed by post-game therapy, or more if there's a serious injury. Eighteen-hour days in season are not unusual and the normal staff of two only has thirty-six of those, let alone any time with their families or, you know, sleep.

Even a team not willing to change drastically could simply hire an extra athletic trainer or two. Every team has a good one at

Triple A and more across the minor leagues. Call one up and the cost is about that of a seventh-round draft pick and the results are more sure and more immediate. Yet, we see baseball teams fail to do this, losing those good workers to jobs that frankly pay better, which surprises people.

More Injury Costs

One of the issues that is huge for those inside baseball and almost completely ignored outside of baseball is a cost center that medical staffs around baseball fear: workers' compensation costs. You probably didn't even think about something like a baseball team needing the same sort of insurance as the nearby Amazon warehouse, but baseball teams are odd businesses. They don't have a lot of employees, though they have more than they did just a decade ago. The players also occupy an odd space, but with workers' comp, teams are dealing with complex state laws.

In fact, it's one state—California, with its worker-friendly and long-lasting laws—that most teams fear most. Keeping their workers' comp premiums low means avoiding injuries as much as possible and minimizing them where they can, but few teams actually do much in terms of real focus on this.

Twenty years ago, baseball's injury consultant produced a book each year that talked about what was good, what was bad, and what could be looked at. MLB changed and stopped publishing it because "no one listened," I was told at the time. (The owner of that firm now raises wagyu cattle, which is probably some kind of metaphor in itself.)

There's an interesting balance in the number of serious injuries, days missed, and dollars lost over the last twenty years. While the dollars have increased along with salary, they've stayed relatively steady when adjusted for inflation. Days lost are up significantly, but at the same time, the serious injury numbers are down, aside from arm injuries.

My theory is that in baseball—and sports medicine as a whole—we've gotten significantly better at dealing with injuries, but that the bigger, stronger, faster mentality, plus the increase in velocity, plus the increases in forces, would have sent injuries up higher had we not gotten better. In essence, sports medicine is controlling the rise, rather than being able to actually pull the number down.

Part of that is the lack of focus on this. Teams that have tried have seen significant decreases. When the Tampa Bay Devil Rays' (at the time) previous owner noticed how much money he was spending on injured players (after signing a number of older players like Fred McGriff and Wade Boggs at the end of their careers), he went and got Dr. Andrews to look at the program. Today, many of the people from that initial change are around baseball—Ron Porterfield of the Dodgers, Ken Crenshaw with the Diamondbacks, Nick Paparesta with the Athletics, and more, and in each of those stops, they've made a big difference.

One European soccer club had an internal rule—for every ten euros they spent on player salary and transfer cost, they should spend one on keeping that player healthy, whether that was sports medicine, sports science, or research. Just their research budget alone was 10 million euros in a season. In MLB, thirty teams together plus the league don't spend that. Most have almost a zero research budget, if not actually zero.

Gary McCoy showed very clearly that a strong, laserlike focus on a program rooted in science was the way forward. He has the results to prove it. With the good people around baseball and on the fringes ready to push the game forward, all they need are the resources and the proper systems.

12

THE CHEATS

In 2021, midway through the season, every baseball fan learned what Spider Tack was and how it had come into the game. MLB, in an unprecedented mid-season crackdown on a rule that had been, at best, lightly enforced prior to this, went as far as patting down pitchers before every inning, looking for any substance that would help a pitcher grip the ball better.

As I talked about in Chapter 1, the ball as manufactured is slick, with a sheen that is taken off with a special mud. Pitchers being pitchers, they have long used extra substances to make the ball do things that they couldn't do naturally. Sticky substances are actually one of the newer variants on this strategy.

Prior to 1920, there were no limitations on what a pitcher could do to the ball and remember, the ball was used differently than today. Balls were returned to play and used over the course of multiple games at times. Home runs were rare. Pitchers used this to their advantage, loading up the ball with all manner of things from tobacco juice to simple mud, used to unbalance the ball, but nothing worked like good old-fashioned spit. Ed Walsh absolutely dominated the American League with his spitball, dancing and diving and called a "freak pitch" by no less than Ty Cobb.

Today, the spitball is not just illegal but a lost art. The last real use of it was Gaylord Perry in the 1970s, though Perry tended to use substances like Vaseline to get the ball to slip.

But what is a spitball, a shine ball, or a mud ball? These dark arts are just another way of making the ball spin, or not, which is how we make any pitch effective. A mud ball is the easiest to understand. A substance is added to one side of the ball, which unbalances the object and creates additional resistance. The pitcher throws the ball with backspin, as a fastball, with the mud on one side. The ball wobbles, almost always toward the heavier side, and acts like a modern cutter, moving to the unbalanced side. If a right-handed pitcher throws a mud ball, he'd put the mud on the left (toward his own glove) and throw it, hoping that the ball would naturally move away from the plate.

A shine ball is much the same. The pitcher would rub the ball against his uniform until the mud and accumulated dirt would vanish. An announcer once said, we are told, that the ball would shine in the sun, hence the name. It was just another way to unbalance the ball, sort of a variant on the mud ball where motion could be induced by a substance or in this case, the reduction of a substance.

The spitball was a whole different beast and there's simply nothing like it in modern baseball (and not just because it's illegal). The spitball and the knuckleball are cousins at worst and perhaps even more closely related than that. While they are thrown differently, the desired result is the same—a ball with greatly reduced spin, creating a pitch that is at the whim of the wind resistance it faces, darting and diving in a way that no one knows is coming—which is why catchers hate the pitch!

For a spitball—and I'll stick (no pun intended) with that term for any non-knuckle variant of the low-spin pitch, just for ease—a pitcher puts a slick substance on either the ball or his fingers. Post-1920, almost all pitchers had to use it on their fingers, hoping no residue would be noticed by the umpires. As the pitcher throws the ball, he tries to "squeeze" it out of his slickened fingertips, rather than having his fingertip catch the ball and impart spin. If you've ever taken a watermelon seed and shot it out of your fingers at someone, it's much the same thing, just bigger and faster.

There are descriptions of the spitball from Walsh's time that describe it as fast, but remember this is relative. It would certainly be faster than a knuckleball, since it can be thrown with a more natural, fastball-like arm speed. While Walsh and Perry's pitches were likely similar, there's little video of either to assess. The spitball is all but apocryphal at this stage, largely because the game changed. Instead of wanting less spin and velocity, pitchers wanted more of both—a lot more.

We know, at least, that Carl Mays's spitball in 1920 was hard enough to kill. His ball, up and in, hit Ray Chapman on the temple and he collapsed, never regaining consciousness. The spitball was quickly outlawed and the ball, and the game, changed forever. While there's—thankfully—no extant film of Chapman's beaning, we do know that similar skull fractures have occurred more recently, but on batted balls that go into the stands, which led to extending the netting at ballparks after a small child was hit in 2019.

Given that we have some exit velocity data and using other sources, it would take a pitch of about 82 mph to fracture the skull. There's a lot of factors that go into this—where it hits, how it hits (impact angle), whether the batter was turning or moving away, as would be natural, and many other factors, so there's quite a broad range. One thing that's key but unknowable was the hardness of the ball. Again, balls were used much longer ahead of the post-Chapman changes and were thought to be softer. If Mays had a newer ball or one that hadn't been softened by a number of hits, then it may have been harder, on the order of modern balls. The story goes that with fading light, a darkened and used ball, plus Mays's hard spitball, Chapman likely never saw it coming.

There's another interesting anecdote about this episode that tells us a bit more about the spitball itself. Mays, who was very mournful for the incident but always insisted he never had intent to hit, let alone hurt, Chapman, mentioned that the "ball was wet from the light rain." Even a dedicated spitballer like Mays needed to control the moisture to make the pitch do what he wanted.

(Another key fact, suddenly more cogent, is the timing of the ban. America was just coming out of the Spanish flu pandemic.

Given what we've gone through with COVID, imagine the implications of spitting on a ball today!)

Sadly, there have been a number of other cases where people have been injured or even killed. The case of Mike Coolbaugh, a minor-league coach who was killed after being hit by a line drive, led to a rule change requiring base coaches to wear a helmet. This has been adopted at all levels. Another change prompted by injuries was the extension of nets after people and children were hit with high-velocity foul balls. Some of these have been recorded at well over 100 mph off the bat. Add in a number of pitchers that have been hit by comebacks and we do have data, but thankfully no other names to go besides Ray Chapman as victims of the game.

There have been attempts to return the spitball to the game. Commissioner Ford Frick settled for making changes to the mound in the 1950s, but discussed bringing back the spitball, calling it a weapon for pitchers. There's some speculation by historian David Bohmer, formerly of DePauw University, that Frick would have seen some of the pitchers that continued throwing spitballs after the 1921 ban, the so-called "grandfathered" few that could legally continue to throw the pitch. There were 17 pitchers allowed to use the pitch after 1921, including Burleigh Grimes, who lasted until 1934, the latest of the group.

That doesn't mean the pitch died, with players like Preacher Roe, Don Drysdale, and the aforementioned Gaylord Perry keeping it alive. A year after he retired, in a 1955 *Sports Illustrated* piece, Roe detailed how he threw the pitch. Drysdale admitted using oil from his hair to slick his fingers, while Perry's autobiography *Me and the Spitter* detailed how he would hide vaseline on his zipper, knowing umpires were uncomfortable checking!

While there are always rumors that some pitcher or another is loading up the ball, the modern era of automatic pitch tracking is making it harder than ever. A pitch with a significantly reduced spin would be noticed almost immediately in the data, if not in person. Many stadiums now use a connected TrackMan device to put spin rate on the scoreboard while MLB.com uses it inside its streaming scoreboard.

Pitches die out for a couple reasons. Even a rule change didn't kill the spitball, but velocity did. A world where Nolan Ryan and JR Richard threw heat was tough enough, but today, every team has a bunch of Nolan Ryans in the pen, throwing high 90s heat and touching 100, with exploding breaking stuff. The knuckleball works, as would a spitter, but even the splitter's usage is reduced.

The other issue is that most of baseball is taught by word of mouth, descended wisdom. There's always talk of Mariano Rivera showing his cutter grip, or a pitcher sharing his changeup grip. Pitching coaches favor things, but who teaches a knuckler, let alone an illegal spitter? That's what saved the pitch in 1939, when Frank Shellenback became the pitching coach for the St. Louis Browns. He'd played for the Black Sox teams with Eddie Cicotte and Ed Walsh and would teach the pitch to anyone, keeping it alive. By the 1960s, there were a few, like Gaylord Perry, who used the pitch, but afterward, velocity and no Shellenback type to keep it going led to a dry period. It probably won't come back, either, leaving it like a hidden trick, something Houdini did once but that no one dares try again.

More Ways to Cheat

There's another variety of illegal pitch that's different from the spitball. The scuff, cut, or emery ball is a variant of the mud ball, where a pitcher uses something to roughen or tear the surface of the ball, which changes the airflow over that side of the ball. The difference in airflow acts much in the same way as the mud ball, but opposite. Where the ball will tail to the mud (weight), it will tail away from the scuff, so the pitcher has to hold it in the opposite way. A right-handed pitcher puts the scuff to his outside, or arm side, in order to make the ball move away from a right-handed batter. Again, this is the rough equivalent of a modern cutter and in some extreme cases, can look more like a slider.

Phil Niekro is one of the last pitchers to be caught throwing a scuff. He could place the scuff to make his knuckleball move against the rough surface. Place the scuff up and the ball would tend to push down, where he wanted it. Niekro was usually very good at hiding his scuffs, but he was caught twice and umpires found an emery board on him once. He later told people inside the game that he'd been advised that the opposing dugout was going to ask for a check. He didn't use an emery board on the field, obviously, so it was misdirection. He was caught and suspended, but they didn't know that he'd replaced some of the stitching inside his glove with a lace that was wrapped in a rough, sandpaper-like surface!

Don't confuse cutting the ball with the pitch popularly known as a "cutter." This is literal—with this, the pitcher or more often the catcher will have a sharp implement, like a tack or nail, hidden on their person. A recent use of this was with a current major-league catcher, who had his glove made with a small flap of leather near the wrist-hole of the glove. He could place a small tack into the flap, which would keep the sharp end away from his skin and toward where he would take the ball out of the glove. He could simply slide the ball down, put enough force on it to cut the surface, and throw it back to the pitcher.

The pitcher then uses the cut as he would a scuff. The ball reacts against the surface and moves, forcing the pitcher to be able to use that motion. In essence, the pitcher and catcher would have to know the next pitch coming before they cut the ball. A cut doesn't help on a fastball or change, but it does on a breaking ball. The ball seldom stayed in play for more than a couple pitches, between foul balls, a ball in the dirt, or even changing after one went in play.

These are hardly the only ways to alter a baseball. After a century and a half, there's still something new under the sun when it comes to pitching, whether it's a grip, a movement, or altering the ball to make it cater to your whims, like a magician. That's really what pitchers are, but sometimes, it is a trick.

Which brings us to the modern fight against sticky substances. It's the opposite of the spitball, which sought to limit spin. Instead, the modern focus on velocity and movement, combined with new tools that measured spin, put a new emphasis on increasing the spin rate. It is also easy to teach, especially if you're willing to "cheat." I put that in quotes because drawing the line on what is cheating and what isn't can be very confusing.

The simplest way to gain spin is to get a better grip. The use of grip enhancers is very simple and has been used for years. There's a rosin bag behind the mound at every game. It was likely an accident the first time a player used a little rosin and noticed then when it mixed with his sunscreen, it got a little sticky. I imagine he got one of those sneaky grins, like the Grinch when he was plotting to steal presents from Whoville. Instead, word got around and everyone did it for years.

But once spin rates became measurable and everything was spin rate and velocity or even a combination—yes, there's a stat for that—that became the easiest element to improve. While velocity is difficult to improve and one gets small increments over time in most cases, spin was much easier to figure out. When rosin and sunscreen wasn't enough, pitchers got creative.

Somewhere along the lines, someone watched a strongman competition. You know the ones, usually late at night on ESPN2 where a bunch of big burly men, usually from Iceland, pull trucks, squat a cheerleading squad, and then lift a series of round boulders up onto pedestals. The guy that played The Mountain on Game of Thrones is one of these guys and at 6 foot 8 and over 400 pounds of muscle, Hafthor Julius Bjornsson can frankly do anything he wants. You tell him no.

Those stones are called Atlas stones and have long been one of the competitions. They look primal and ancient, but you can actually buy your own stones for under a thousand bucks. At 100, 150, and 200 pounds, the shipping might be a bit steep and your FedEx guy is going to hate you.

In strongman competitions, those stones are heavier. Competitors pick up four stones and place them on pedestals, with the largest stone weighing 440 pounds! The record for this is a tick over 16 seconds. (Yes, there's video and it's amazing.)

However, picking up hundreds of pounds of round stones is about more than just lifting. It's about grip, too, and while rosin is not in short supply, almost all elite competitors use something called Spider Tack. Like the Atlas stones, it comes in three grades, ranging from light—described as "thin and pliable"—to heavy—"webby and very viscous." Again, it's cheap (twenty bucks for a tube), effective, and easier on your delivery man.

If it works for helping to lift a 400-pound rock, you can bet it works to keep a pitcher's fingers on a ball, spinning out at the last second and adding as much as 500 rpm to a pitch. When sticky stuff was pulled out of the game in mid-2021, pitchers in the week prior to the enforcement of the long-existing rules were seeing 300, 400, and more rpm reductions. I can't prove it's because they stopped using Spider Tack or something similar, but it's not just a coincidence.

Of course, pitchers aren't going to just quit cold turkey. Some pitchers got creative again and this time, someone's mom does some quilting. A substance—again, easily found on Amazon—called basting spray gave pitchers a perfect combo. The substance is sticky, but easily removed without a trace. Even better, it's non-toxic, so a pitcher can lick his fingers, which is often a tell for a pitcher using something stickier. Basting spray doesn't stain, so it can go on fabric as it's designed to, and is removed with alcohol. In a 2021 dugout, a bottle of Purell is never far away, and no ump is going to take that away.

Given all the ways that pitchers can cheat, does it actually help? Harold Mozingo, a former minor-league pitcher and now coach, loves to experiment. He couldn't afford a force plate for his facility, so he hacked a Wii board to measure his pitchers. When he started working on spin, simple sticky stuff wasn't going to do and he started experimenting, checking his results against a Rapsodo. Pine tar gave him an extra 200 rpm, with the longtime sunscreen/rosin mix well behind at 130. What was best? A Super Blow Pop, that childhood treat—and yes, available on Amazon—turns out an amazing 700 rpm.

A recent experiment by Travis Sawchik, a writer at the Score who helped write the bestselling *The MVP Machine*, found that Spider Tack added 700 rpm to the baseline, though he didn't test Blow Pops. In multiple tests, which admittedly aren't journal-quality studies, it's clear that sticky substances help and that more stick helps more.

Rosin. Spider Tack. Blow Pops. Basting spray. Is there nothing pitchers won't use, even with the threat of a ten-game suspension?

As a quirk, one of the rules of baseball expressly states that a pitcher can't use a foreign substance on the ball to alter it. One of the substances called out is licorice. I talked to everyone I knew, from historians to pitching coaches, and no one had any idea why it would prove so advantageous.

Naturally, MLB historian John Thorn had the answer, providing me with an article that quoted a player named Tom Seaton. Seaton discovered that licorice turned his saliva black. He'd rub half the ball with the now-black saliva and then throw a fastball. He said it was "some fooler, I tell you!"

I'd love to know how much licorice five cents got you, or how Seaton's hypno-licorice pitch actually worked. (A black and white ball? That's testable.) Seaton was pitching for the Brooklyn Tip-Tops and the Newark Peppers back then, part of the breakaway

Federal League, but he did have 27 wins for the Phillies in 1913, which even for that era was pretty good. He led the league in strikeouts and walks that season, so that licorice pitch must have been something.

The Art of the Steal

In a chapter about cheating, we simply can't avoid sign stealing. You probably immediately thought "Astros" and that's okay, but sign stealing is universal in baseball. Don Zimmer was a good player and a manager and coach in MLB for over fifty years and to a man, people still point to Zimmer as the best sign stealer in the game. Zimmer would sit on the bench or lean over the rail and watch the coaches go through their signs. With his years of experience and some innate skill, he could discern signs quickly. There were teams against which he didn't try it, knowing that they knew he was watching, so they would change signs every inning or put in a more complex system.

It's not just Zimmer, though. Watch almost any player when he gets on second base and you'll likely see some not very subtle sign stealing. The runner can see the signs and the catcher and will do something to pass on as much information as possible. If the catcher sets up inside, the runner might put more weight on that foot. He can watch the pitcher's grip and have his hands in or out, for example, to indicate breaking ball.

Teams at all levels have similar systems and look for tells from pitchers. If a fielder is setting up in a certain place, expect a fastball in. If the pitcher's glove shifts, he's throwing the slider. All of these are in the scope of the game. It's cheating by the rulebook, but it's accepted and encouraged in the real scope of things. There's even a bit of unwritten rules about it, like telling your opponent if you're not going to face him anymore. When I was coaching in college, we'd regularly exchange a note with the opposing team like, "Sir, your reliever there in the fourth is tipping his curve." Usually, the batters have already told him.

So what did the Astros do that went beyond the pale? First, I'll acknowledge that it's hardly just the Astros that have amped things up beyond the long-accepted levels of sign stealing. The Blue Jays had a system that used people wearing different colored shirts in the outfield seats. Allegedly, a person behind them had binoculars and could tell white or blue in front of him to stand up and give the indication. Other teams have used cameras, something done even at the college level.

The Astros used a series of cameras, but took it even a step or ten further. Not only would they collect the video, but they used machine-learning systems to try and decode signs. Yes, this is what Don Zimmer did in his head, but the Astros could do it more quickly and at an even higher success rate. They would match signs to plays and pitches, giving them a database that went beyond a simple play-by-play, making the sign stealing even more effective. Whether they used buzzers or trash cans to pass on the information, the weak link was clearly getting the info to the players, and their hubris in doing it in such inelegant ways is what got them noticed and then caught.

Cloud computing is advertised as part of MLB's statistical packages, first AWS and now Google Cloud, but the Astros took it a step further. They could gear up the equivalent of thousands of computers in the cloud and break down patterns they saw in near real-time. Something they saw at first pitch could be analyzed against data sets, making the system smarter and smarter, faster and faster.

The downside to this is that by pushing the limits so much, the Astros have likely set back similar uses by years. Machine learning, visual-based data sets, and neural networks are all technologies that could help baseball teams, but are now a bit darkened by a poor first use.

Drug Testing

In 2005, I wrote a book called *The Juice*, and while performance-enhancing drugs are no longer the open issue that they were in baseball for the previous decade (and longer, if we're going to be honest), they still exist in the game. The idea that BALCO or Biogenesis, run by hucksters selling pseudoscience, were created in a vacuum is ridiculous. People knew which clubhouse attendants could get good drugs and which could get good seats at a restaurant and that's still the case.

In 2021, there are people getting caught in MLB, but it's often for the use of old-school steroids, the ones from the '80s with six-month detectable periods. MLB's testing program has proven to be remarkably effective at both offering enough of a penalty to keep most from even trying and at catching those that aren't deterred. (Most pro sports do one or the other, or barely exist, such as the NFL's testing regimen.)

Drug testing is similar to the way in which the cops on *Law & Order* make the victim look through page after page of mug shots. If the perp's not there, they won't be identified. The reason that the BALCO users weren't caught initially is that the drug didn't have a "mug shot"—a profile that the tests were looking to match against. When a rival coach turned in the drug itself, the testers had a match and a test for it in weeks.

Since Patrick Arnold found his drug in an old East German textbook, others have been looking for old or new drugs that can slip by the system and as yet, there have only been a few attempts and most of those were close enough to other drugs that the metabolites were identical. Busted.

The other way of trying to beat a test is to come in low, under the threshold of what is considered a positive test. Instead of taking a lot of steroids in a cycle, athletes from Biogenesis were taking small doses in very controlled patterns. The downside was, it didn't work. It was, at best, pseudoscience from the Boschs and while some of their customers didn't test positive, it's because of testing schedules, not undetectability.

There is a third way and with this, there is almost no way to detect the doping. That is, using bioidentical substances, which naturally occur in the body. The key anabolic agent used today in multiple sports is insulin. The same drug that helps diabetics lead a normal life can be used to make huge anabolic gains. It's effective, cheap, readily available, easy to transport, and undetectable. The downside is that if one gets the dosing wrong, they'll die. It's a risk some players are willing to take.

In terms of new substances, almost all the research has shifted to supplements. An entire subculture has shifted from steroids to ketones, including many of the people from baseball's so-called steroid era. These substances are legal and effective, but also require strict ketogenic diets that can be very difficult to maintain. Exogenous ketone esters like D-BHB, acetoacetone, and others can be very effective, but are also relatively new, so long-term effects aren't known. Early attempts were a bit rough, with one researcher describing the taste as "jet fuel," but results have been notable. It's significant that ketone esters seem to act on blood glucose, much in the same way that insulin does, so these could be working in the same pathways.

Baseball has, in the space of twenty years and multiple iterations, essentially driven steroids out of the game. They're still used at the fringes by the desperate or misguided, but those are largely caught. Absent a major new anabolic steroid being discovered, there's only one path where baseball and all sport once again fall behind.

That's in genetic manipulation. When I first wrote about the possibility in 2005, it was theoretical. Tools like CRISPR were years away but were anticipated even by the World Anti-Doping Agency (WADA) in the makeup of what they call the "Biological Passport." With this, the body is checked for normal levels of many natural substances. An increase from those levels would set off a warning. This is being done in Olympic sports, where it has controversially been used against several runners with naturally high testosterone, but this is a naturally high level rather than a change created by doping.

While genetic manipulation hasn't gotten to a point in humans where specific areas like muscle growth, speed, or eyesight can be manipulated, the technology exists that could do such things in the not so distant future.

I once stood on the field at PacBell Park—now Oracle Park in San Francisco. I was there to meet Barry Bonds for an interview and I watched him take batting practice on the field. At the same time, Neifi Perez walked out and started taking grounders at shortstop. He put on a show, fielding every ball, flipping it behind his back, slapping it freehand to the second baseman, and even somersaulting and flipping the ball jai alai style to first. It was the best display of fielding skill I'd ever seen. Another player walked up next to me and must have seen the look on my face. "First time seeing Neifi?" he asked. I nodded.

"Just remember," he said, leaning forward. "Neifi really sucks, man. I don't know how he's still on the team."

13

ANALYTICS

"I know perfectly well that baseball cannot be played one
hundred percent according to figures, and that the human
element is even more important. I realize that certain sets
of figures on players and teams will change from time to
time, but nevertheless, by a deep and systematic research
into the detailed statistics which I have in mind, there is
bound to come to light numerous facts which were previously
unknown, and which would prove of great value."

– Allan Roth

There are few words that split a room of baseball fans down the
middle like "analytics." There is also no concept that has brought
more change to the game than the numbers and formulas that have
accompanied its rise in the sport.

We all know the story of Bill James, "The Sultan of Stats," who
landed his first job in baseball at the ripe age of 53 after years of
working odd jobs; or Billy Beane, whose Oakland Athletics teams
were built on the framework of James's brainchild—sabermetrics—to propel them to multiple, consecutive playoff appearances
despite a bottom-five payroll. But how many people know the story
of Allan Roth, baseball's first true statistician?

Abraham "Allan" Roth, a tailor from Canada, got his first job
in baseball with the Brooklyn Dodgers in 1947. The first man in
baseball paid by a team as a statistician, it was Roth who saw value
in player splits, spray charts, and statistical breakdowns. These
concepts seem mundane and old hat to us today, but when Roth

presented these numbers to general manager Branch Rickey, it was an untapped well of new ideas.

For the next eighteen years, Roth recorded every pitch thrown in a Dodgers game. His handwritten 17x14-inch papers were then circulated throughout the clubhouse and used to craft the most optimal lineups. "Baseball is a game of percentages," said Roth.. "I try to find the actual percentage, which is constantly shifting, and apply it to the situation where it will do the most good."

Roth believed that runs batted in was an overrated stat, unless directly correlated with the chances to drive them in, and impacted by which base they'd been driven home from. It was his analytics that convinced manager Burt Shotton to move Jackie Robinson, with only 12 home runs at the time, to the cleanup spot in 1949. The cleanup spot had traditionally been reserved for the lineup's best power hitter, but despite finishing the season with only 16 home runs, Robinson knocked in 124 runs on his way to winning the league's Most Valuable Player.

We can attribute the creation of saves, isolated power, and on-base percentage to the work of Allan Roth. The metrics he created provided the game with new methods of player evaluation that are still in use today: on-base percentage is and will perhaps always be one of the most important statistics in sabermetrics.

Today, all thirty major-league clubs have analytics departments made up of mathematicians, scientists, and free-thinkers made from the same mold as Allan Roth. It was his blueprint that continues to lead to the creation of new equations and theories, new tools of measurement for success, and paradigm shifts in the game of baseball. "He was the guy who began it all," says Bill James. "He took statisticians into a brave new world."

Prior to the birth of analytics and the dedicated statisticians who tout it, baseball statistics were more limited. But numbers have always been the official "currency" of baseball since Harry Chadwick invented the box score in 1859. Averages, hits, runs, strikeouts ... things we could see from our seats were the end-all be-all measurements for players. But it was Allan Roth, and more famously, Bill James, the Sultan of Stats, who found a way to add quantifiable numbers to assess things that previously could only be scouted with "the eye test."

James coined the term "sabermetrics," a portmanteau of metrics and SABR, the Society for American Baseball Research, in 1971 while working the night shift at a Kansas pork and beans cannery. This terminology describes a whole swath of recently created, complex statistics developed by him and other mathematicians, opening a locked door to a room full of previously never-before-contemplated information.

We can thank James for creating a number of numerical innovations, like range factor and defensive efficiency rating: two statistics that describe a defender's capabilities on the field. Some say that these two equations alone make both fielding percentage and "the eye test" obsolete.

Nevertheless, many talent evaluators and front-office executives look at sabermetrics skeptically to this day. Despite the success of the early 2000's Oakland Athletics documented in Michael Lewis's bestselling book, *Moneyball*, not everyone has jumped aboard the analytics train.

I spoke with Ari Kaplan, an innovator and pioneer of the sabermetrics movement. Kaplan was one of the first statisticians to introduce sabermetrics into a major-league front office, and one of the most successful to do it. Tasked with creating the analytics department for the Chicago Cubs, Ari has dealt with his fair share of believers and non-believers.

Although he describes his early years working for the Dodgers (his first MLB job) as "extremely positive," Kaplan admits that over the years in his work with other teams he's faced pushback from stakeholders who not only rejected sabermetrics, but showed "reluctance to even meet to determine what it (was) they are rejecting."

This sentiment illustrates much of baseball's unwillingness to change, even if it is for the better. Change is never immediate and usually faces opposition from traditionalists, both inside the game and outside, watching from the executive suites or the stands.

While statistics like on-base percentage and WHIP (walks plus hits per inning pitched) are still core tenets of sabermetrics, many of these numbers and principles have fallen out of favor over the years as the game has evolved and new metrics are created.

WAR (wins above replacement) is one such metric that has been devalued and its importance and relevance debated by experts. Simply put, WAR is the number attached to a player representing how many wins he's personally responsible for. It is calculated by combining batting runs, base-running runs, fielding runs, positional adjustment, league adjustment, and replacement runs, then dividing that by runs per win. Seems pretty simple, right? But as sabermetrics have evolved, it has convoluted the value of WAR and necessitated changes in how it is used.

First, WAR does not take into account late-inning production or critical situations in close games. Nor does it account for the defensive metrics used by the infield shift, another advancement of sabermetrics, which muddies up DRS (defensive runs saved), inflating WAR for infielders who happen to be positioned in the right place. Also complicating WAR is the concept of "the opener," a reliever who starts a game, only to pitch an inning, maybe two. Figuring out how to classify the opener, whether as a starter or reliever, based on innings pitched and other factors, is still being worked out by sabermetricians, and therefore, so is WAR.

While WAR may have lost some of its prestige and is being adjusted and rethought to include these concepts, several metrics that were once popular have lost favor amongst sabermetricians, becoming obsolete and abandoned.

But how does a prominent metric become obsolete? I asked sabermetrician Keith Woolner, a former Baseball Prospectus writer, now principal data scientist for the Cleveland Guardians, whose creation of VORP (value over replacement player) while at BP replaced TPR (total player rating) as one of the key concepts of player evaluation.

"A metric becomes obsolete when people stop using it," says Woolner. "It's really nothing more than that. If you can convince enough people to get behind your metric (especially if some of them are influential), other people start being interested in it, having other authors use it in articles, research, etc. then other alternative metrics may tend to lose mind share. Ideally, that would happen because there's something 'better' in the metric replacing them. That could be using a better source of data, developing a more accurate model, needing fewer assumptions, simplicity of presentation, catchier naming, or ease of use because sites you use have computed it for you. Some of it is no doubt attributable to better 'marketing' of a metric through frequent communication and persistent promotion, too."

Here, Woolner illustrates the volatility and ephemeral nature of any metric. It does not take much for something to become obsolete in an age when more and more information becomes available all the time. And as Woolner points out, there is always a need for new metrics, especially as the game continues to evolve. And much of baseball's evolution is the result of the ubiquity of sabermetrics in baseball.

Every action has a reaction, and for every strategic move employed there is a countermove from the opposition. For example, the infield shift has created the need for several hitters to start focusing on hitting more fly balls. The increase in fly balls has led to more pitchers throwing high, hard fastballs to counteract this trend, which in turn has sent the major-league strikeout per plate appearance to an all-time high (a whopping 23.4 percent!!) in 2020.

More strikeouts mean more pitches thrown, and that causes longer games, which Rob Manfred has taken upon himself to shorten. He's even induced the minor leagues to experiment with a cap on pick-off attempts and a ban on the infield shift. If these rule changes are determined to be "successful" and make their way into the majors, how much of an effect will it have? Will sabermetricians make stolen bases en vogue again once a pitcher has expended all of his throws over to first base, giving the runner the opportunity to take a giant lead?

It remains to be seen how the game will change and how saber-metrics will then also have to change to remain current. Some sabermetricians feel the next step in combating the record-high fly-ball rates is to employ four-man outfields. If the infield shift is banned, what will analytics say about countering that?

"As long as research continues, new metrics will be churned out, and some of them might stick" says Woolner. "The potential rule changes MLB is experimenting with could change how we view some long-familiar stats."

We're still awaiting the verdict on these rule changes, but per-haps much of this development will come through the advancement of technology. The Statcast system, installed in every big-league park in 2015, captures data like exit velocity, launch angle, spin rate, and even a defender's route efficiency. Using the TrackMan component of Statcast, sabermetricians have found that pitches with a lot of movement are more effective than high velocity.

This optical technology has advanced biomechanics and artifi-cial intelligence in such a way that you can break down a player's body into little parts in order to analyze every bit of movement. For Ari Kaplan, an innovator and pioneer of the sabermetrics movement, capturing as much in-game data as we are able is vital to the goal of enhancing our ability to make as many informed predictions as possible.

"Artificial intelligence is key to look at the complex interactions of a variety of data," says Kaplan. "For example, with using data to analyze a swing, you would want to understand what components of the swing (hand, shoulder, waist, legs) are most effective for each individual player, and relay if they change something in their swing will it be helpful or just push a weakness to another part of their swing."

On Opening Day 2020, Major League Baseball debuted Hawk-Eye, the first of its kind in artificial intelligence, compiling and measuring said data not only to improve on-field production, but also to study the body more closely for technique. Hawk-Eye uses its twelve 4K cameras capable of capturing 120 frames per second to break down the body into nineteen points to show limb and torque movements, whereas before we just looked at one central data point.

Prior to Hawk-Eye's implementation, there was no apparatus for tracking the movement of the bat or following the gyro-spin, the three-dimensional spin of the ball. TrackMan's system could not track the seam of the ball, meaning that spin rate was simply inferred. Thanks to Hawk-Eye, we've opened up all new realms of detailed biomechanical analysis for hitting, pitching, and fielding.

As data-driven technology continues to advance, so will sabermetrics. And as sabermetrics advances, baseball will continue to change. Old metrics will become obsolete as new ones are created to keep up with the game. Batting average, hits, home runs—they will always be important, and nothing will change that. But sabermetricians have created new statistics for player evaluation that have permanently altered our appraisal of those numbers.

We will continue to see sabermetrics progress and values change as technology evolves the game. I know a lot of fans think analytics have become too intrusive, and perhaps that's true. The debate within the stadium walls will continue to rage as well between "old school" and "new school."

The beauty of baseball and technology's partnership is the capacity to see things that could not be seen before. The "eye test" just no longer cuts it. New metrics are being created and used, ushering us into a new age of scouting and number analysis.

We can put together terabytes of information on a single player and use it as a predictor of their success or to identify any weakness. It truly is amazing what we've been able to accomplish so far, but there is still more to come.

The days of Harry Chadwick's newspaper box scores as the end-all be-all of player evaluation is long gone. The numbers have grown tenfold in quantity and in their complexity. Allan Roth's spray charts and pitch selection, once recorded on looseleaf paper in a seat by the dugout, are now documented by computers placed all over the stadium.

These same computers then create a detailed spreadsheet of all said players' tendencies: what a pitcher throws in a certain count to left- or right-handed hitters, where a batter hits a certain pitch and when. Technology has truly taken over a large part of the analytics portion of the game. But the things Roth believed were important then, still are. They're just analyzed through a different lens (no pun intended).

Analytics through technology have also had a large impact on the body and injury prevention. Through Hawk-Eye, coaches and doctors can evaluate every tiny movement in a player's body and determine whether said player is at risk of an injury. Kinesiology, the study of body mechanics and movement, is more prevalent then ever today and used by many front offices in deciding whether someone's prior injury history presents a reasonably expected risk of recurrence for the particular athlete or if it's just an aberration, and he is worth taking a chance on.

But analytics are not perfect, much like any science. New statistics are often fine-tuned for years until they are deemed adequate while old ones, held in high esteem for years, are suddenly meaningless and superannuated. That is certainly a result of big tech getting exponentially better, but also due to adjustments teams make to the adjustments, constantly fine-tuning to find another fractional advantage.

Front offices in tandem with managers and coaches have created spreadsheets that tell their infielders exactly where to position themselves in the infield, outfielders in the outfield, and the pitchers in some cases have an entire at-bat's worth of pitches planned out ahead of time.

In these ways, baseball has changed and become more scientifically and mathematically inclined. Blue-blooded Ivy Leaguer statisticians have become prevalent in many front offices, contrasted with prior years where front offices were all ex-players and scouts. Analytics and technology have opened up new worlds for baseball, yielding a much different result-oriented sport than we've seen in its history.

Despite the backlash from traditionalists, sabermetrics' biggest metric is its prominence and acceptance in today's game. Even the casual fan understands the impact of analytics and the changes it's brought with it. Who would have thought we would see Statcast numbers shown and discussed on a national telecast, when just a few years ago, the crew at Baseball Prospectus was being told to keep their nerd stuff to themselves?

Just as Tony La Russa can go from groundbreaker to outdated in a decade, baseball can change as well when it comes to digging deep into the sabermetrics side. Analytics and science have made both parts of the game and many of its constituents obsolete. But that's what change does.

So how will technology and analytics continue to change baseball? Well, for starters, double-A baseball tried out a new rule in 2021 that would put a premium on an infield's ability to shift by forcing all infielders to stay on, or in front of, the dirt. This is one of Rob Manfred's experiments to combat and minimize the impact of analytics on the game. He has heard the calls from a growing dissatisfied group of fans who feel that baseball has changed for the worse due to the increased presence of science and technology in the game.

Would banning the shift in the majors lead to more (groundball) hits, fewer strikeouts, and fewer flyballs? Yes. Would it make the game more interesting to restore it to some of its past un-analytically driven glory? Maybe not, but it's still an interesting idea. I'm not convinced that baseball has lost a whole lot of fans and star power since analytically thinking front offices have become the norm. Unlike the NBA, where after five minutes of viewing it you can complain and declare confidently that "all they do is take three-pointers now," the integrity of the game of baseball has not changed...just the priorities.

Analytics has changed the way the game is played on the field like it has the NBA. Granted, baseball still uses the same age-old principles to get on base and score, while an NBA offense looks so strange comparatively nowadays, as every ball rotation is made with the intention of freeing up an outside shooter for three. A center in today's game is more of a deterrent for any player trying to drive to the basket and score, rather than a scorer himself like in the day of Kareem Abdul-Jabbar and Moses Malone. But that's what analytics and science have taught us about basketball: a robust three-point percentage is the key to winning a game. As a result, that's how NBA team rotations and lineups are constructed.

Although similarly in baseball, the starting lineup has also changed in how we prioritize each spot in the order. Thanks to analytics, the leadoff spot is no longer for the fastest hitter (although it is often still the premier spot for them). The second spot is not just for the best pure hitter on the team, and the third and fourth slots don't have to be reserved for the best two power hitters on the team.

Number analysis shows us that the earlier spots in the order receive more at-bats per game, which has led managers to have typical cleanup hitters batting leadoff and second. Baseball has become very home run–oriented, so banking on your best power hitters to hit them early and often is now the preferred strategy for scoring runs. Often managers will structure their lineup around generating the most runs using the metric of weighted runs created plus (wRC+), changing around the 1–8 hitters on a nightly basis depending on the pitcher, ballpark, and other external factors.

That is what wRC+ really is, taking the metric of "runs created" and incorporating external factors like the ballpark or the pitcher's tendencies. Here is what that looks like:

((wRAA per PA + league runs per PA) + (league runs per PA - ballpark factor x league runs per PA) / league wRC per plate appearance, not including pitchers) x 100.

Essentially, what this equation does is give us a chance to compare players who play in different ballparks and for different periods of time. For example, two players with the exact same stats will have different wrC+ if one plays in hitter-friendly Coors Field and the other puts up the same numbers in Petco Park, a notoriously pitcher-friendly stadium.

WRC+ is just another of the fancy new methods managers employ to get an edge on the opposition. Look at that equation and tell me that's not science!

Again, this all could change if Rob Manfred deems his experiment successful and bans the infield shift, leading to a different approach at the plate for hitters who no longer have to hit one over an outfielder's head to get on base.

Another concept that has been re-thought under analytics/sabermetrics is bunting. In 2011, Bill James told an interviewer that bunting is "usually a waste of time."

"I mean, if you think about it, [a] bunt is the only play in baseball that both sides applaud," James said. "The home team applauds because they get an out, and the other team applauds because they get a base. So what does that tell you? Nobody's really winning here." Not coincidentally, the number of bunts in the league went way down after that and has yet to come back up.

I find this interesting considering the amount of space infielders give hitters when they shift. These last few years I've seen holes on one side of the infield so large, that if the batter laid down a bunt, he could arrive at second base before anyone picked up the ball. But unfortunately (or fortunately for the analytically inclined), bunting is becoming less and less a part of the game everyday.

You know what else is becoming less and less a part of the game everyday? Starters going deep into games. Yes, that's analytics too. Studies have shown that a starting pitcher's effectiveness begins to wane as he approaches 100 pitches and/or the third time through the lineup, and managers have taken note. Baseball Prospectus refers to it as the "Times Through the Order Penalty."

It is worth noting that when any starting pitcher is going through the order a fourth time, that pitcher is still in the game because he's throwing well. Any pitcher who isn't performing well enough to get through the lineup three times is not going to be out there the fourth time around.

These numbers show why a starting pitcher generally doesn't approach the complete-game threshold, and help explain why bullpens have changed so much through the lens of analytics and science.

It is not uncommon to see teams fill their 25-man rosters with 12 or 13 pitchers nowadays. Thanks to analytics, bullpens have become much more specialized and require more bodies to fill the specific rolls meted out to them. This is known as "bullpen slotting," the act of designating each and every reliever to a certain roll. Closers, setup men, lefty-specialists—every reliever is labeled and expected to step onto the mound with a particular purpose.

Managers seem to follow scripts with their bullpens in a way where you can almost predict who will be coming in in a certain situation. Especially once the starter hits 100 pitches, you almost expect him to be removed from the game at the next available, logical juncture. Gone are the days of the firemen: relievers like

Mike Marshall, Rollie Fingers, or Kent Tekulve who often appeared in 80–100+ games and were used at any time or for any purpose. Analytically speaking, today the recipe for a successful relief pitcher hinges on his ability to do just one thing well—not several—and for a shorter time. Those three pitchers I named didn't just get into a lot of games, but once they took the mound, they were in there for multiple innings. That's another thing we just don't see much these days in the era of the specialist.

With today's "bullpen games" becoming more and more popular, we see managers using several of their relievers early in the game, constantly switching them out in an attempt to keep the opposing team's starting lineup from getting into a groove.

While analytics have made labels more pertinent and germane to relief pitchers finding success in the league, it has also devalued the closer spot in the bullpen. It is no surprise that the top 12 relief pitchers in yearly salary in the 2021 season were all closers (or were closers when they signed their contract, as in the case of Zach Britton). Closers are the recipients of the fanciest of all the statistics for a relief pitcher: saves. Saves clearly lead to bigger contracts, but how is a "save" any different from the "hold" the inning prior? Technically it's not, but saves mark the end of the game and the closer gets to look like a hero for putting a game in the win column.

The Leverage Index, according to MLB.com, "measures the importance of a particular event by quantifying the extent to which win probability could change on said event with 1.0 representing a neutral situation." So what we have here is a metric that says that a pitcher coming in with the bases loaded, down by three runs will be above 1.0, meanwhile a pitcher starting the 9th inning with no men on and up by three runs will register below 1.0.

This lets us know that very often the highest leverage situations in any given ballgame don't belong to the closer. Protecting a one-run lead in the 9th is certainly high leverage, but a closer rarely ever is called to pitch in the middle of an inning like the setup or specialist pitchers. Closers come in with a clean slate and finish off the game with fewer variables than the pitchers who preceded him. Relievers coming in with inherited runners face higher-leverage situations than most closers do on a nightly basis and get a fraction of the fanfare for it (and a fraction of the paycheck).

Watching a baseball game from start to finish does seem more formulaic today, and we can thank analytics for that. The completion of a full nine-inning ballgame does at times feel like a science. So much more goes into building a starting lineup or a roster than ever before. With so many metrics being used to find new ways to win ballgames and labels being attributed to players, it has become even more imperative to find the right guys for the right spot.

It is also the job of the front office and team sabermetricians to examine ways to combat other teams' decision-making. This has led to what some see as scripted games. Analytics has become so pervasive in the sport that it dictates what any given game is going to look like before the players have taken the field. Every infielder knows where he's supposed to be for every batter, and pitchers often know every pitch and in what order they will attack a specific hitter. But of course, every player still has to execute their plan, and that will never be a sure thing.

Baseball is and has always been a game of adjustments, and teams who can make said adjustments on the fly on a consistent basis are still the teams who are going to be the most successful. Analytics has just changed the way we analyze and respond to these adjustments, but the principle is still the same. Baseball has become more scientific for sure, but that has not made it any less fun or interesting.

14

THE FUTURE

In 2000, Keith Woolner—then with Baseball Prospectus and now with the Cleveland ball club for over a decade—wrote a paper based on the Hilbert Questions, a fin de siècle challenge to mathematicians, in which he details twenty-three problems he felt had yet to be adequately addressed by the then-nascent field of sabermetrics. Two decades later, over half of Woolner's questions remain unanswered, but he's working on those questions as principal data scientist for the Cleveland Guardians.

A couple years later, when the original essay from the Baseball Prospectus book was reprinted on the website, another writer added a couple of his own questions in a 2004 article. James Click asked some great questions then, and he's now the general manager of the Houston Astros.

The lesson here is clearly that asking good questions can lead to some big jobs.

The future then is twofold. We have to answer questions that we've been asking for years without an adequate answer and we will have to ask new questions as the game and the technologies around it continue to advance.

Woolner didn't mention things like spin rate because while everyone knew the ball spun, no one knew how important it was and there was no way to quantify it easily until the invention of the TrackMan. Click didn't question the survivability of 100 mph pitchers because we had only seen glimpses of it in outliers like Nolan Ryan. Aroldis Chapman's MLB debut was four years in the future from Click's article.

I know that the best way to look like a fool is to make predictions, so I will close with some new questions of my own, though

I will openly admit that some of these were first articulated to me by either people around the game or readers of my previous work.

These are my version of new Hilbert Questions. It is a challenge to baseball as a sport and to those who love it. I give a hearty tip of a Cleveland "Block C" cap to Woolner for the idea. I hope that there's someone picking this book up twenty or thirty years from now that will look at these and think, "What? We didn't always know this?" the way that you or I do about something like spin rate or launch angle.

How do we assess the value (or reduced value) of a manager on a game?

How do we quantify the real effect of injuries? (Before you ask, I do not believe injuries can be predicted, even twenty-five years from now. I hope I'm wrong.)

How do we quantify the value of non-baseball personnel, like scouts, analysts, and general managers?

Can we better assess the value of players in terms of actual dollars earned, in comparison to not just some theoretical construct like WAR, but in actual dollars received by a team?

Will legalized gambling, including microgambling on a pitch-by-pitch basis, lead to better analytics or put baseball in danger of a second Black Sox scandal?

Can we better quantify a model for player development?

Can we assess player potential by non-baseball means, such as genetic testing or functional assessments?

Is it possible to create baseball players from outside the pyramidal baseball system? (Example: Is the next Nolan Ryan in Mozambique and if so, how do we find him? Also known as the Tom House Million-Dollar Conundrum.)

Can we redefine the aging curve to better account for early success and later success?

Can we more accurately assess the likely success of a stolen-base attempt predictively, or will we continue to rely on guesswork and tradition?

Is it possible to determine the maximum velocity with which a human can throw a baseball under normal conditions and if so, is there also a maximum survivable velocity? (I'm calling this the Chapman Conundrum.)

Is there a "calling game" skill, and how should it be credited—to the catcher or coach calling the game, or for the pitcher who is executing the pitches? (This one is the Codify Conundrum.)

Is variance from projections an inaccuracy of the model, or can we discern real value, positive and negative, of an individual from the delta off a largely accurate predictive model (the Pecota Conundrum)?

My hope is that this book and these questions will challenge you and that in twenty years, we'll know all these answers. But what will baseball be like in twenty years? There's no better way to be wrong than to make predictions, but I think I have to here. Let's discuss short-, medium-, and long-term trends.

In the short term, I think that machine learning and AI will find a legal place in the baseball lineup. Tools like StatStak will help players translate data into actionable intelligence more easily, by "speaking baseball" and using graphs and video to help make mountains of data more understandable.

I think that robot umps are not only inevitable, but should be welcomed in the areas where they can improve the game, which is balls and strikes to start. I think we'll always need men at the bases, and I'm curious where the home-plate ump will stand if he's not calling balls or strikes. Does it buzz down to him what to call, which would make the game look traditional in terms of how we're used to seeing them, both in person and on television? Add the fact that at lower levels, I think those technologies will come in more quickly than most think as camera systems get better and cheaper.

Fan experience is also going to change. As ticket prices continue to increase, as well as ancillaries like concessions and parking, fans will demand a better experience. They'll want more personalized experiences, delivery of food, and great views. They'll use their phones or video glasses to get stats, see replays, or listen to mic'd up players on the field.

I think we're going to see an evolution of defense, whether or not the shift is banned or made tougher. The science of fielding is just in its infancy and ways to get better are going to adjust to any rule.

In the medium term, I think we're going to continue to see change. As Terence Mann, the James Earl Jones character in *Field of Dreams*, says, the only constant in America is baseball. It's a great line, but down to the baseball itself, it's not actually that constant. Baseball changes with America and with the world, evolving alongside it. It seems like it's always been the same, while what we call constant is actually just its presence in our lives.

I think one of the biggest changes in sport will be the creation of a holistic approach to athlete health management. The last fifty years have been a hodgepodge of learning, with advancements in sports medicine and sports science unevenly distributed with practitioners trying their best to keep up and to keep their athletes healthy under changing circumstances. We're nearing a point where sensors, cameras, wearable technologies, and the uber-connected "Internet of Things" becomes a reality that starts paying off for athletes.

I've seen early iterations of several systems of what I'll call an "AthleteOS," ranging from the very simple to the medical grade. There will be fits and starts, with baseball likely staying behind sports like European football and the NBA in terms of being cutting edge. The advantage that baseball has is the sheer number of games, at all levels. One-hundred sixty-two is a big sample size, especially when broken down to individual player responses. I think this will happen inside of ten years and be absolutely ubiquitous to the point of not even being thought of as an oddity by the twenty-year point.

Equipment changes are likely. Who knows how MLB will resolve its ball problem—Dr. Meredith Wills is young—but we're likely to see advances. As climate change becomes more of an issue, bats will need to change, perhaps to an engineered wood first, which could create problems as that manufacturing is put in place. Hot bats could cause a peak in homers and exit velocity.

I also think that in the next twenty years, we'll see the first women players at upper levels and, I hope, the major leagues. I don't think that will come from softball conversions, but girls that fight to play and stand their ground. I do think it will take something quirky—a submarine pitcher, a Jamaican speedster who becomes a pinch runner, or a rangy defensive specialist—but that's just to start. Just as Roger Bannister ran the four-minute mile and seemed to make it possible for everyone else, that first Jacqueline Robinson will break down walls all over again.

In the longer term, I think baseball is going to be just fine. Imagine being like Steve Rogers, waking up sixty years down the line. Just like Rogers would have seen differences between the game of his day—Ted Williams and Bob Feller among them—there would be more similarities than not. I think it's just as likely that fifty years from now, we'll walk into a baseball stadium and it will be more familiar than not, at least on the field.

We'll likely be amazed. In the same way our great-grandfathers would be stunned at high-def video screens and $18 beers at Dodger Stadium, we'll be stunned at whatever comes along in terms of the in-stadium look and feel. The athletes will be uniformly bigger and stronger and that will come with both positives and negatives. My hope is that we'll have better protective equipment and way better sports medicine and sports science to keep them on the field.

What I don't think we'll have is robots in the field. Maybe umps, but I think this is a human game and fifty years won't be enough. If robots are made for any game, it's American football anyway.

A friend recently emailed me and said she just didn't care about her team. They were losing, hard to root for, traded away their best players as they tanked, and generally didn't seem concerned that she canceled her season tickets the previous year. I get it. Sometimes our love for our team makes it hard to root for them. Look at the number of people on Twitter or other social media where their handle or avatar is their team. They could present themselves any way they want to the world, and here they are, defining themselves as a fan of a team.

Baseball is far from perfect. It's a provincial game, not always welcoming and certainly not logical. You can fall in love with the game all over again and then fall out of love just as easily. Twenty years ago, Ichiro electrified the game, an improbable imp of a baseball player, slashing and running his way into an all-time great and a true baseball character. As I write this, Shohei Ohtani is leading the majors in homers, all while being a top pitcher, breaking WAR charts. Maybe the next superstar is playing catch on a field in Tokyo, falling in love with the game as well.

The game can be enjoyed in so many ways. I've drilled down into stats and I've casually watched on a hot afternoon in the bleachers, a couple beers in. I've coached youth to college, worked with pro pitchers, and consulted with MLB teams. It's not the same game, really, at all those levels, but it's one of those things where you can find a piece of it that's all yours and you can also just let it wash over you and be entertained.

Too many people will tell you there's only one way to love this game. They'll tell you they know the right way and will whisper the unwritten rules. They might say that analytics are ruining the game or that scientific thought and method have no place in what amounts to a child's game. They'll tell you there's greedy owners ruining the game and players getting overpaid.

Don't listen. Baseball is there and in fifty years, I think it will be there for that generation as well. Love the game the way you want, whether you're Mike Trout or little Cindy Blue, and I hope it loves you back.

ABOUT THE AUTHOR AND ACKNOWLEDGMENTS

Will Carroll writes about injuries. His writing has been featured on ESPN, *Sports Illustrated*, Baseball Prospectus, and the *New York Times*. His work has been recognized by the American Association of Orthopedic Surgeons (AAOS), the Society for American Baseball Research (SABR), and the Fantasy Sports Writers Association (FSWA.) He lives in Indianapolis.

This book would not have been possible without the invaluable research and assistance from David Barshop. David wrote the chapter on analytics and assisted with a number of other projects within the book.

Part of Chapter 8 ran originally on Baseball Prospectus in February 2006. Copyright was retained by the author after publication and runs in an edited and updated format.

Special thanks to everyone at Skyhorse Publishing, especially Julie Ganz. Thanks to Peter Gammons, Dr. Meredith Wills, Keenan Long, Dr. Mike Sonne, Butch Baccala, Gary McCoy, Cory Schwartz, Ken Arneson, Trevor Forde, Joey Stevenson, Sarah Howard, Brittany Dowling, Ben Hansen, Steph Armijo, Alan Jaeger, Xan Barksdale, Ari Kaplan, Mat Kovach, J Daniel, Rob Miller, John Thorn, John Mayer, Joe Sheehan, Caleb Vaughn, Al Ready, Josh Kusnick, Bryan Goelz, Tyler Brooke, and Joel Henard.

And a number of people that I couldn't quote directly in the book for various reasons.

**The author has or had consulting relationships with items and teams mentioned in the book. He was a consultant for ProPlayAI and for one major-league team at the time of writing, while he previously worked at Motus Global. Also, he and Gary McCoy work together at the new sports science startup Northstarr. Their interview happened prior to both joining the company.

**Chicago Style is used in this book, which states that professional titles are not capitalized. For members of the NATA out there, please note.